Mössbauer Effect
Methodology

Volume 7

MÖSSBAUER EFFECT METHODOLOGY

Proceedings of Annual Symposia Sponsored by
the New England Nuclear Corporation, Boston

Edited by Irwin J. Gruverman

Volume 1
Proceedings of the First Symposium–January 1965

Volume 2
Proceedings of the Second Symposium–January 1966

Volume 3
Proceedings of the Third Symposium–January 1967

Volume 4
Proceedings of the Fourth Symposium–January 1968

Volume 5
Proceedings of the Fifth Symposium–February 1969

Volume 6
Proceedings of the Sixth Symposium–January 1970

Volume 7
Proceedings of the Seventh Symposium–January 1971

A Publication of the New England Nuclear Corporation

Mössbauer Effect Methodology

Volume 7

Proceedings of the Seventh Symposium on
Mössbauer Effect Methodology
New York City, January 31, 1971

Edited by

Irwin J. Gruverman

Manager, Nuclides and Sources Division
New England Nuclear Corporation
Billerica, Massachusetts

Ꝑ PLENUM PRESS • NEW YORK–LONDON • 1971

Library of Congress Catalog Card Number 65-21188

ISBN-13: 978-1-4615-9004-0   e-ISBN-13: 978-1-4615-9002-6

DOI: 10.1007/978-1-4615-9002-6

© 1971 New England Nuclear Corporation
Softcover reprint of the hardcover 1st edition 1971

Plenum Press, New York
A Division of Plenum Publishing Corporation
227 West 17th Street, New York, N.Y. 10011

United Kingdom edition published by Plenum Press, London
A Division of Plenum Publishing Company, Ltd.
Davis House (4th Floor), 8 Scrubs Lane, Harlesden, NW10 6SE, London, England

PREFACE

This is the seventh volume of a series that provides a continuing forum for publication of developments in Mössbauer effect methodology and in spectroscopy and its applications.

Mössbauer Effect Methodology, Volume 7, records the proceedings of the Seventh Symposium on Mössbauer Effect Methodology. The symposium was sponsored by the New England Nuclear Corporation and interest was concentrated on spectroscopy, with some attention to methodology and applications. The Symposium was held in the Mercury Ballroom of the New York Hilton on January 31, 1971. Dr. N. Benczer-Koller presided over the afternoon and evening sessions.

About two hundred participants attended, demonstrating the continued high level of effort in the field and of interest in this series. Austin Science Associates, Elron, and Nuclear Science Instruments demonstrated their Mössbauer equipment products. These were well received, and evidently are continuing to improve in utility and quality.

Applications papers reported on local magnetic moment measurement and radiation damage studies. A large number of spectroscopy papers was presented and subject matter included work on Mössbauer parameters in iron, tin, and europium and on conversion electron spectroscopy. Methodology work reported included a source for Sn^{119}-Sb^{121}, a "blackness" distortion removal technique, and use of radio frequency fields. A paper is included on calculational techniques although it was not ready for presentation at the symposium.

This volume marks an experiment in format which departs from that of the previous volumes in this series. The authors were requested to submit their papers in camera-ready form. The advantage of this procedure is unusually swift publication, at the expense of some differences in style among the various papers, expressing the author's individuality rather than an editor's attempt at conformity.

The editor is indebted to his colleagues for their assistance in soliciting and selecting papers. Dr. Benczer-Koller, in addition to her superb management of the lengthy program, was most helpful here. Carl Seidel, Robert McKay, and Nancy Snook all deserve thanks for general assistance, facilities and refreshment organization, and secretarial assistance, respectively.

Since the Symposium, the editor has rejoined New England Nuclear as a full-time employee and looks forward to renewing his relationships with many of the readers of these volumes.

                                                                    I.G.

Needham, Massachusetts
June, 1971

LIST OF CONTRIBUTORS

J. O. Artman, Carnegie Institute of Technology and Mellon Institute of Science, Carnegie-Mellon University, Pittsburgh, Pennsylvania

G. M. Bancroft, Department of Chemistry, The University of Western Ontario, London, Canada

P. H. Barrett, Department of Physics, University of California, Santa Barbara, California

L. H. Bowen, Department of Chemistry, North Carolina State University, Raleigh, North Carolina

A. J. Carty, Department of Chemistry, University of Waterloo, Waterloo, Ontario, Canada

H. Z. Dokuzoguz, Babcock and Wilcox Research Center, Lynchburg, Virginia

B. D. Dunlap, Argonne National Laboratory, Argonne, Illinois

P. A. Flinn, Carnegie Mellon University, Pittsburgh, Pennsylvania

R. B. Frankel, Francis Bitter National Magnet Laboratory, Massachusetts Institute of Technology, Cambridge, Massachusetts

Y. Hazony, School of Engineering and Applied Science and
    Computer Center, Princeton University, Princeton,
    New Jersey

Juergen Heberle, Department of Physics and Astronomy, State
    University of New York, Buffalo, New York

Randolph L. Lambe, Department of Physics, University of
    North Carolina, Chapel Hill, North Carolina

T. K. McNab, Argonne National Laboratories, Argonne,
    Illinois

W. T. Oosterhuis, Physics Department, Carnegie-Mellon
    University, Pittsburgh, Pennsylvania

Loren Pfeiffer, Bell Telephone Laboratories, Incorporated,
    Murray Hill, New Jersey

W. M. Reiff, Department of Chemistry, Northeastern
    University, Boston, Massachusetts

Dietrich Schroeer, Department of Physics, University of
    North Carolina, Chapel Hill, North Carolina

Brian B. Schwartz, Francis Bitter National Magnet
    Laboratory, Massachusetts Institute of Technology,
    Cambridge, Massachusetts

H. D. Sharma, Department of Chemistry, University of
    Waterloo, Waterloo, Ontario, Canada

Charles D. Spencer, Department of Physics, University of
    North Carolina, Chapel Hill, North Carolina

Jon J. Spijkerman, Scientific Research Corporation, Alva,
    Oklahoma

H. H. Stadelmaier, Department of Engineering Research,
    North Carolina State University, Raleigh, North
    Carolina

K. A. Taylor, Department of Chemistry, North Carolina State
    University, Raleigh, North Carolina

M. Celia Dibar Ure, Carnegie Mellon University,
    Pittsburgh, Pennsylvania

P. G. L. Williams, Department of Chemistry, The University
    of Western Ontario, London, Canada

CONTENTS

METHODOLOGY

APPLICATIONS

USES OF THE MÖSSBAUER EFFECT IN RADIATION-DAMAGE STUDIES

Dietrich Schroeer, Randolph L. Lambe and
Charles D. Spencer

Department of Physics, University of North
Carolina at Chapel Hill, North Carolina

ABSTRACT

*A review is presented of contributions which Mössbauer
spectroscopy can make to studies of radiation-damage effects.
Special emphasis is placed on the unique aspects of this
technique, including the following examples: in all Coulomb
excitation, as well as in some radioactive decay processes,
every Mössbauer gamma ray has its origin in a radiation-damage
event, leading to high sensitivity; all chemical charge states
of a Mössbauer ion can be simultaneously observed and identified through the isomer shift; the Mössbauer effect is particularly sensitive to the local details of the damage effects;
and radiation-damage processes with recovery times as short as
the nuclear lifetimes can be studied. Some of the more significant contributions of the Mössbauer effect to a variety of
radiation-damage questions are illustrated by reviewing experiments from the existing literature.*

I. INTRODUCTION

In this paper we would like to review the contribution
which the Mössbauer effect (ME) can make to radiation-damage
(RD) studies. A quotation from the introduction to Crawford
and Billington's book RADIATION DAMAGE IN SOLIDS [1] summarizes the main reason for the present widespread interest in
RD studies:

...fast-particle bombardment affords a means of controlled

3

introduction of lattice defects into crystalline solids
and hence is an excellent tool for studying in a general
way the influence of lattice defects on the physical properties of solids.

Until now, the ME has been little applied to RD studies.
This is suprising since it can give some unique information
about the induced lattice defects. One unique aspect is the
high sensitivity for RD found in two types of experiments.
Particularly appealing is the case where RD occurs in an excitation process such as Coulomb excitation (CE) or in some radioactive decays; since then every emitted Mössbauer gamma ray
has its origin in a RD event. One might call these one-to-
one experiments. The other case occurs when studying RD effects in a radioactive source. Then the Mössbauer ion is an
impurity in the lattice as the parent and/or the daughter
element. Since such an impurity ion may have a special attraction for the induced defects, large RD effects may be
observed by ME spectroscopy. These might be called defect-
coupled experiments. Another unique aspect of the ME in RD is
its ability, through such parameters as the isomer shift, to
both uniquely identify and quantitatively measure the various
resulting chemical charge states of the Mössbauer ion. A
further important aspect is the possibility, when the excitation of the Mössbauer level and RD event are coupled, to study
RD which is as shortlived as the Mössbauer nuclear state itself.

We will first discuss the various possible RD processes
and the classes of defects induced by different types of irradiation. Then we will look at all of the Mössbauer parameters and consider what specific information each might
yield about the induced defects. We will illustrate each
case by considering experiments from the existing literature.

II. NATURE OF DEFECTS PRODUCED BY IRRADIATION

The types of defects produced in a lattice depend very
strongly on the nature of the irradiation [1,2]. The usual
definition of RD implies that an actual displacement of atoms
takes place in the lattice, though there may also occur under
irradiation noticable changes in the lattice through displacement of electrons. If the incident bombarding particle can
transfer about 25 eV to a lattice atom, then this atom can be
displaced from its lattice site and RD occurs. Otherwise ionization is the probable mode of energy loss, and the resulting

lattice modifications such as color centers do not constitute RD. We will now briefly discuss the types of defects produced in the lattice under the different types of irradiation.

A. Bombardment by Neutrons and Nuclei

For nucleons and nuclei a bombarding energy of 1.5 keV is sufficient to insure that even the heaviest atom can be given the 25 eV of recoil energy required for a displacement in the lattice. Thus RD displacement processes are quite likely during reactor activation or excitation [e.g. Ref. 3], in cyclotron activation, in Van de Graaff Coulomb excitation [e.g. Ref. 4], in ion implantation [e.g. Ref. 5], and in fast-particle bombardment of absorbers [e.g. Ref. 6]. Figure 1

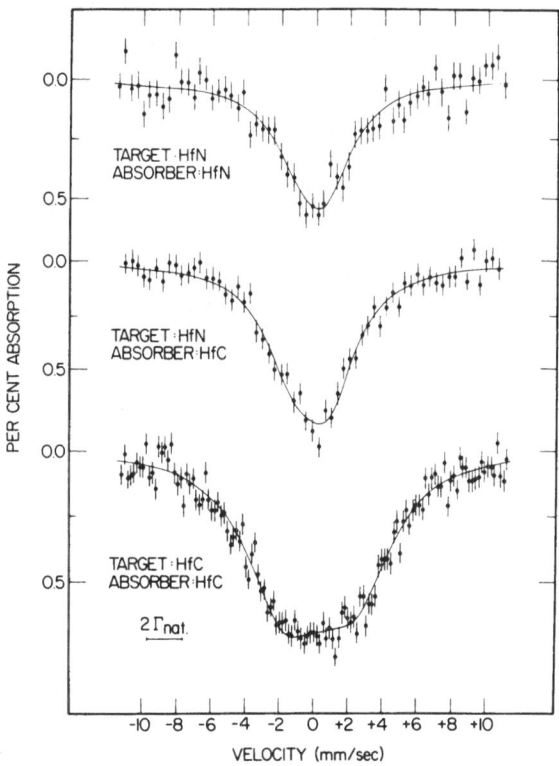

Fig. 1 Mössbauer spectra of Coulomb-excited ^{178}HfC and natural HfN targets. (After Fig. 1 of Ref. 4.)

illustrates a one-to-one effect for CE [4]. When used as a
Mössbauer absorber, HfC shows a reasonably narrow resonance
line, whereas when used as the target for CE the emission
spectrum shows an increased quadrupole splitting; indicating
that the displaced Hf atom has come to rest in distorted lattice surroundings.

B. Bombardment by Energetic Electrons

The defects induced in a lattice by energetic electrons,
as from a 2 MeV Van de Graaff accelerator for example, are
basically similar to those induced by fast particles. This
can be seen in Fig. 2, which shows that the Mössbauer magnetic

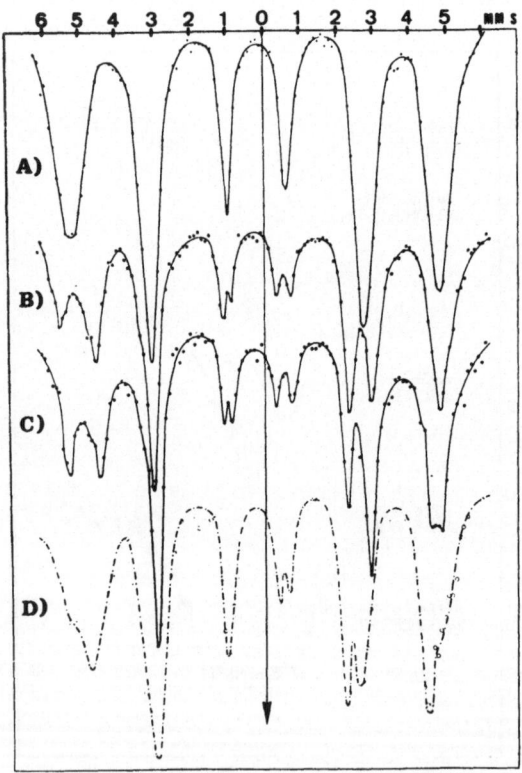

Fig. 2. Mössbauer absorber spectra of Fe-Ni alloy samples:
A) non-irradiated polycrystals, B) a single crystal irradiat-
ed by neutrons, C) a single crystal irradiated by electrons,
D) polycrystals irradiated by electrons. (After Fig.2 of Ref.6)

hyperfine structure (HFS) in an Fe-Ni alloy absorber after
neutron and electron irradiation is essentially the same.
There is, of course, the disadvantage in electron bombardment
that no one-to-one excitation experiments are possible. However, a compensating advantage is provided by the fact that
to displace atoms in a lattice by electrons requires energies
ranging from 150 keV for oxygen to about 1 MeV for rare earths
[e.g. Refs. 1 and 2]. This means that by choosing lower energy electrons for bombardments it is possible to selectively
displace only atoms with low atomic weights, thus giving further information about the actual displacement processes.

C. Bombardment by High-Energy Gamma Rays

Irradiation with energetic gamma rays is typically carried out in a "bomb" containing a many-Curie source of ^{137}Cs
or ^{60}Co. The damage processes for such gamma rays with energies on the order of 1 MeV are primarily due to indirect interactions via energetic Compton electrons [e.g. Refs. 1 and
2]. The resulting defects will therefore be similar to those
produced by bombardment with electrons. One advantage of using gamma rays is their high penetrating power, making this
technique particularly useful for irradiating bulky samples,
or samples involving encapsulated radioactive sources which
cannot be exposed to electron bombardment. Figure 3 illustrates these possibilities. The three spectra are those of a

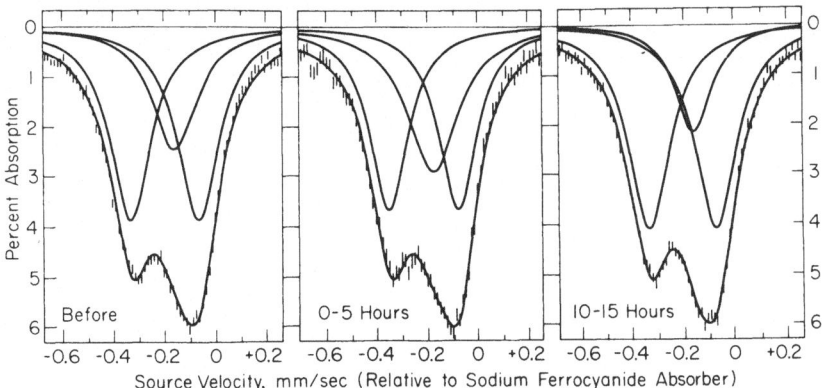

Fig. 3. Mössbauer source spectra of ^{57}Co in Co_3O_4. Each
spectrum is fitted with a quadrupole-split doublet and a singlet: a) prior to irradiation, b) 0 to 5 hours after irradiation with 661-keV gamma rays, and c) 10 to 15 hours after
irradiation. (After Ref. 7.)

sealed $^{57}Co_3O_4$ source before, shortly after and half a day
after irradiation in a ^{137}Cs bomb [7]. The changes in the
relative intensities of the doublet clearly indicate not only
an irradiation effect but also show the relaxation of this
effect on a time scale of several hours. In this case the
observed effects indicate a mixture of physical lattice RD
and of rearrangements in the electron shell of the decaying
Co atom.

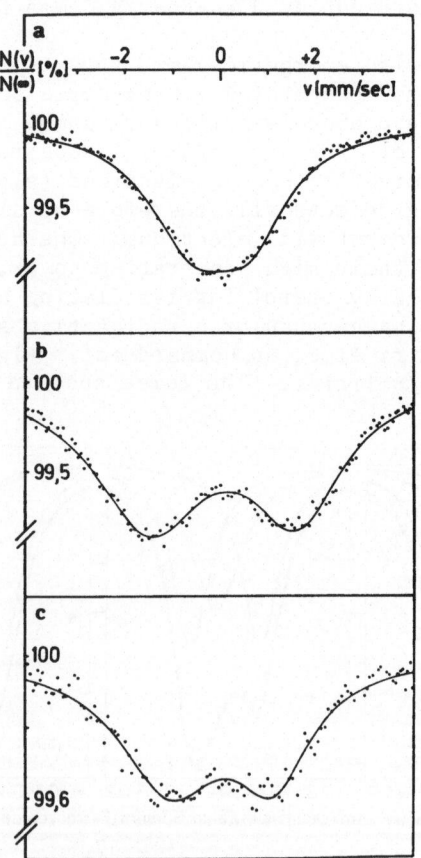

Fig. 4 Mössbauer spectra of gadolinium-oxalate absorbers:
a) unirradiated, b) after exposure to unfiltered Hg light,
c) after irradiation with x rays. (After Fig. 5 of Ref. 9.)

D. Bombardment by X Rays

X rays are generally too low in energy to produce atomic displacements either through direct collisions or through the collision of a scattered photon or Compton electron. One exception is the production of Frenkel-pair defects in alkali-halides [e.g. Ref. 8] through processes referred to as radio-lysis. But generally one expects x-ray irradiation to result only in electronic bond rearrangements [e.g. Ref. 9]. For example, Figure 4 shows that the x-ray produced quadrupole splittings in the Mössbauer absorber spectrum of gadolinium oxalate resembles that produced by unfiltered Hg light. Therefore x-ray irradiation is likely to be primarily useful in RD studies by indicating which effects observed in high-energy irradiations are due to ionization processes rather than due to RD. See Refs. 10 for similar non-RD effects in irradiation experiments.

E. Radiation Damage by Radioactive Decay

In radioactive decay processes the recoil of the decaying atom is of primary importance. In alpha decay, for example, one would expect a displacement of the emitting atom to occur most of the time [e.g. Ref. 11]; while in beta and K-capture decays, the probability of displacement is controlled by the energy distribution of the electron or neutrino spectrum, and hence should usually be low. Figure 5 shows the Mössbauer spectra of CoF_2 doped both with radioactive ^{57}Co and with stable ^{57}Fe [Ref. 12]. There clearly are some aftereffects from the low-energy K-capture decay visible in the source

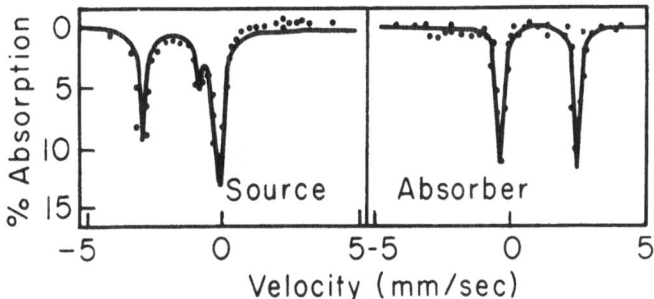

Fig. 5. Mössbauer spectra of CoF_2 doped with both ^{57}Co and 1/2% ^{57}Fe. The source spectrum shows Fe^{3+} lines which are not present in the absorber spectrum. (After Fig. 2 of Ref. 12.)

spectrum, effects which result in some of the daughter iron
ions being for some time in the trivalent state. In agreement
with the usual interpretation of such phenomena [e.g. Refs.
12-14], this large effect is probably due to ionization processes which lead to strong coupling of the impurity ion to
some already existing defect; although it conceivably might
also be due to a radiolysis process such as displacement of
the momentarily highly ionized ion due to Coulomb repulsion
[e.g. Ref. 15].

III. MÖSSBAUER PARAMETERS IN RADIATION DAMAGE

We will now illustrate some of the specific effects of
defects on various parameters in Mössbauer spectra, and examine the significance of each particular parameter. (See Ref.
16 for a parallel discussion for defects induced by techniques
other than RD.)

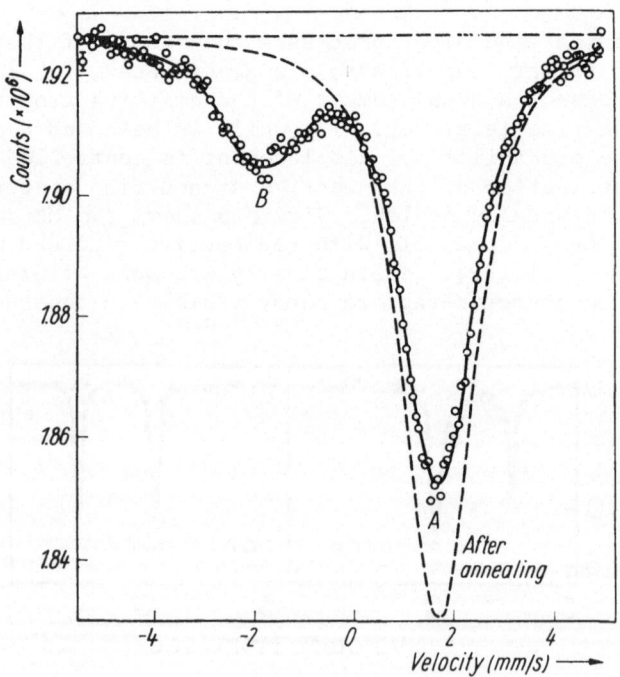

Fig. 6. Mössbauer source spectrum of neutron-irradiated
Mg$_2$SnO$_4$. Line A corresponds to the "normal" tetravalent
state, while line B corresponds to divalent tin. (After
Fig. 4 of Ref. 18.)

A. Isomer Shift

One of the most attractive aspects of Mössbauer spectroscopy is the fact that for many of the Mössbauer transitions
the different ionic charge states can be clearly distinguished
and characterized by their isomer shifts. Figure 6 illustrates such a characterization. It shows the source spectrum
of Mg_2SnO_4 following neutron irradiation [17,18]. The divalent state induced by the irradiation is clearly distinct
from the normal tetravalent state. After annealing, the spectrum shows the recovery of the central tetravalent line at the
expense of the divalent line. Clearly Mössbauer spectroscopy
makes it possible not only to distinguish the various charge
states, but also to measure their relative and/or absolute
intensities.

Similar RD effects have been reported for neutron irradiation in other tin compounds [19-21]. Berger, et al. [22]
found that the reaction $^{56}Fe(n,\gamma)^{57}Fe$ in $FeSO_4 \cdot 7H_2O$ apparently
led to about 40% of a trivalent singlet as well as the normal
quadrupole-split doublet. Apparent covalency changes have
been observed in Eu_2O_3 during Coulomb-excitation [23,24], in
PbTe under neutron irradiation [25], and in Fe-Al alloys under
neutron irradiation [26]. In general these examples illustrate that the largest effects are observed in strongly-
coupled and in one-to-one excitation experiments.

B. Magnetic Hyperfine Interaction

The magnetic hyperfine interaction is less sensitive to
RD than the isomer shift. This is due to the fact that the
magnitude of the HFS for a given charge state is primarily
dependent on long-range order, and hence responds only mod-
erately to local changes in the lattice structure. Of course
a change in charge state will be reflected in the HFS, but
this is largely a duplication of the isomer shift information.
One must conclude that primary information from the HFS in RD
studies is probably in the effects of local order-disorder
changes. This is illustrated by the Mössbauer spectra in
Fig. 7 of Fe_3Al both as an absorber and as the target for a
neutron-excitation study. Czjzek and Berger [3] deduced
from these spectra that the displaced and excited iron atoms
come to rest in a lattice site with nearby random order in the
lattice. This is demonstrated in the figure by the fact that
the two distinct magnetic HFS for the absorber have become a
random average of HFS in the excitation experiment. Berger,
et al. [22] found for the same reaction that in $\alpha - Fe_2O_3$ the

magnetic field at the iron nucleus decreased by about 4%.

Gross changes in the magnetic structure of absorbers, as in the magnetic transition temperatures, might be detectable. For example, Gros and Pauleve [6] were able to extract, from the HFS of Fig. 2, information about the various types of ordered domains produced by neutron and electron bombardment in an Fe-Ni alloy. In the Coulomb excitation of ^{73}Ge in Cr the magnetic order was apparently destroyed by the irradiation [5].

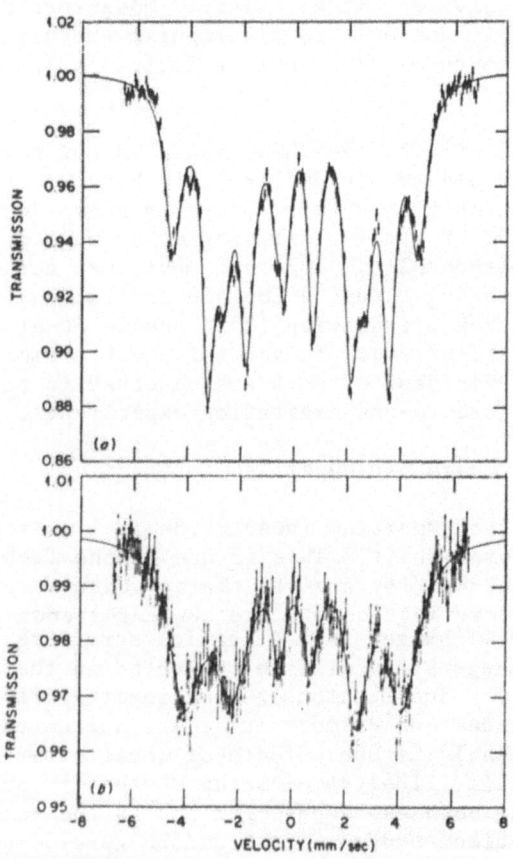

Fig. 7. Magnetic HFS of Fe$_3$Al used a) as absorber and b) as target for the ^{56}Fe(n,γ)^{57}Fe reaction. (a) can be fitted with two six-line magnetic spectra, while (b) is fitted with six independent Lorentzian curves. (After Fig. 3 of Ref. 3.)

C. Electric Quadrupole Interaction

The electric-quadrupole interaction in Mössbauer spectroscopy is somewhat more sensitive to local lattice structure, making it a promising parameter for RD studies. The
largest effects occur if the charge state of the Mössbauer
atom changes, so that the Sternheimer factors amplify the
crystalline field differently, or if the atom is displaced
from its normal lattice site into one with a different symmetry. Both of these effects will be particularly prominent
in one-to-one and defect coupled experiments where only displaced Mössbauer atoms contribute to the observed spectra.
This was illustrated in Fig. 1 for CE. Similar effects have
been observed in tungsten carbide [27] as well as in other
hafnium compounds [28]. Quadrupole-splitting changes associated with charge state changes have been seen in neutron irradiations of various tin compounds [18-21]; gross changes in
electric field gradients occurred for overall neutron and
electron irradiations of Fe-Ni absorbers as indicated in
Fig. 2 [6]. Quadrupole doublets have also been observed in
implanatation studies, for example when [57]Co is implanted into
diamond [e.g. Ref. 29]; and in neutron irradiation of silicon
doped with [57]Co [30].

D. Magnitude of the Mössbauer Effect

One of the first observed cases of RD in ME spectra was a
reduction by a factor of 4 in the magnitude of the [237]Np
Mössbauer resonance following the alpha decay of [241]Am over
that observed after the beta decay of [237]U [11]. Since then,
such reductions have been frequently observed following CE
[6,31-34] and neutron excitation [35,36]. The usual interpretation is that some fraction of the Mössbauer atoms are
distributed over so many different lattice sites and defect
structures that their Mössbauer spectra simply form a diffuse
background to the distinct resonances from the more normally
situated Mössbauer atoms. However, the possibility of a thermal heat spike cannot be excluded, and in some cases it may be
simply a matter of a locally reduced Debye temperature due to
lattice defects near the Mössbauer atom. A change in the
Debye temperature was put forward, to explain the reduction in
the Mössbauer f factor observed for [197]Au sources produced
by neutron irradiation at 4°K in Pt foils [3]. Figure 8 shows
the resonant fraction as a function of irradiation time for
this case.

The magnitude of the ME observed in excitation experiments
helps us to understand a little better the details of the RD
process under bombardment by fast particles. Theoretical calculations by Vineyard, et al. [38-40] and by Dederichs, et al.
[41] indicate that, for metals at least, most of the collisions should be replacement collisions with the displaced
atom coming to rest at an ordinary lattice site with perhaps
some vacancies nearby. If primarily interstitial atoms were
produced in RD processes, then no ME would be visible because
of a smearing of the different spectra corresponding to the
many possible lattice sites; the typically small reduction in
the f factor in CE supports this replacement model. The observation of the ME in these one-to-one experiments also indicates that, in agreement with calculations [42-44], any heat
spikes in the lattice must essentially have died out within
the lifetime of the nuclear Mössbauer level.

E. Other Experimental Parameters

In Mössbauer RD studies there are several variables a-
vailable in addition to the above Mössbauer parameters. As
indicated above, these variables include the control of the
irradiation conditions such as the kind of bombarding particle
[e.g. Figs. 2 and 8], the intensity and dosage of the irradiation [e.g. Fig. 8], and the irradiating temperature; as well
as control of the temperature following irradiation to induce
annealing. Figure 6 showed an anneal for Mg_2SnO_4 following
neutron irradiation; Fig. 9 presents a more detailed record of

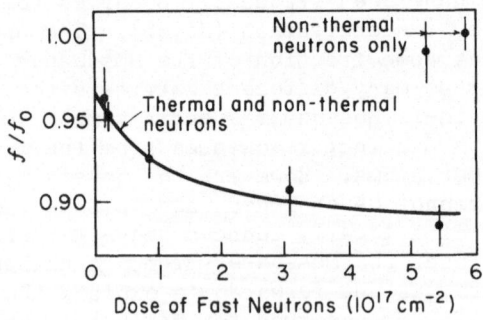

Fig. 8. Reduction of the Debye-Waller f factor for ^{197}Au
produced by neutron irradiation at 4°K in Pt. (After Fig. 5
of Ref. 37.)

this anneal. From the annealing rate as a function of the
annealing temperature, one can extract the activation energy
of the particular defect.

Relaxation of the RD effects can be expected to occur
following any irradiation process. In Fig. 3 there are clearly visible changes in the Mössbauer spectra of ^{57}Co in Co_3O_4
induced by gamma-ray irradiation, changes which relax with
lifetimes of several hours. Similar effects have been observed for the isomer shift in PbTe [25]. If the defect and
nuclear lifetimes are comparable, then relaxation effects such
as line broadening or mixtures of charge states might be directly visible in the Mössbauer spectra. For several years
the broad Fe^{3+} lines in ^{57}Co-doped non-stoichiometric oxides
were thought to represent such relaxation phenomena [e.g.
Ref. 13]. References 45-47 give some cases where unusual
charge states or lattice sites were observed in the decay of
^{161}Tb to ^{161}Dy, representing possible relaxation effects.

Finally we suggest that excitation and radioactive-decay
studies are, in one sense, equivalent to producing RD by means
of a pulsed source, with a pulse repetition rate equal to the
nuclear lifetime of the relevant Mössbauer nuclear state [see
Ref. 48 for comparison]. This means that very short-lived defect states with lifetimes on the order of nanoseconds can be
studied, states which otherwise might never be observed. By
properly selecting the measurement temperature for example, it
may even be possible to determine the lifetime of such states

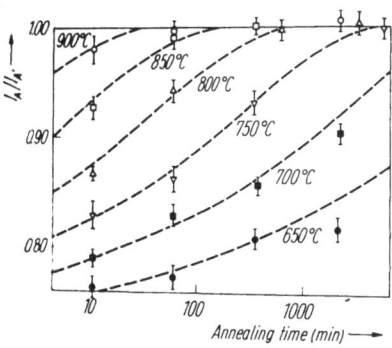

Fig. 9. Isothermal annealing data for the intensity I_A' of
the normal Sn^{4+} line in neutron irradiated Mg_2SnO_4. The
dashed lines represent a theoretical fit assuming a discrete
defect activation energy of 3.5 eV. (After Fig. 10 of Ref. 18.)

through the detection of relaxation phenomena. Such "pulse" techniques open up some previously unexplored time ranges in defect phenomena.

IV. SUMMARY

While the possibilities of using the Mössbauer effect in radiation-damage studies have not been much explored to date, they do seem quite promising. There are unique opportunities in excitation and radioactive-decay aftereffect studies to explore the local nature of defects, and to explore these defects on a very short time scale. Through one-to-one effects, and through the coupling of impurity atoms with specific defects, very large-scale RD-induced variations in Mössbauer parameters can be observed. Up to now the ME has been found useful in measuring quantitively all ionic charge states simultaneously; and in studying RD replacement probabilities and heat spikes in heavy particle bombardment.

V. ACKNOWLEDGEMENTS

This work has been supported by the Materials Research Center of the University of North Carolina under contract SD-100 with the Advanced Research Projects Agency and by the Air Force Office of Scientific Research, USAF, under Grant No. AFOSR-69-1716. The United States Government is authorized to reproduce and distribute reprints for governmental purposes notwithstanding any copyright notation hereon. We would like to acknowledge helpful conversations with John P. Stott.

VI. REFERENCES

1. D. S. Billington and J. H. Crawford, Jr., RADIATION DAMAGE IN SOLIDS, Princeton, N. J.: Princeton University Press, 1961.
2. L. T. Chadderton, RADIATION DAMAGE IN CRYSTALS, N. Y.; John Wiley, 1965; G. J. Dienes and G. H. Vineyard, RADIATION EFFECTS IN SOLIDS, N. Y.: Interscience Publ. 1957; and A. C. Damask and G. J. Dienes, POINT DEFECTS IN METALS, N. Y.: Gordon and Breach, 1963.
3. G. Czjzek and W. G. Berger, Phys. Rev. B1, 957 (1970).

4. C. G. Jacobs, Jr., N. Hershkowitz, and J. B. Jeffries, Phys. Letters 29A, 498 (1969).
5. G. Czjzek, J. L. C. Ford, Jr., J. C. Love, F. E. Obenshain, and H. H. F. Wegener, Phys. Rev. 174, 331 (1968).
6. Y. Gros and J. Pauleve, J. de Physique 31, 459 (1970).
7. C. D. Spencer and D. Schroeer, Bull. Am. Phys. Soc. 16, 24 (1971); and to be published.
8. J. H. Crawford, Jr., Adv. in Physics 17, 93 (1968).
9. W. Heidrich, Z. Physik 230, 418 (1970).
10. A. Yu. Aleksandrov, N. N. Delyagin, K. P. Mitrofanov, L. S. Polak, and V. S. Shpinel, Sov. Phys.-JETP 16, 1467 (1963), a translation of ZhETF 43, 2074 (1962); N. Saito, H. Sano, T. Tominaga, and F. Ambe, Bull. Chem. Soc. Japan 38, 681 (1965); and V. I. Goldanskii, THE MÖSSBAUER EFFECT AND ITS APPLICATIONS IN CHEMISTRY, N. Y.: Consultant's Bureau, 1964, p. 42.
11. J. A. Stone and W. L. Pillinger, Phys. Rev. Letters 13, 200 (1964).
12. J. F. Cavanagh, phys. stat. sol. 36, 657 (1969).
13. G. K. Wertheim, Phys. Rev. 124, 764 (1961); W. Triftshäuser and P. P. Craig, Phys. Rev. 162, 274 (1967); and H. N. Ok and J. G. Mullen, Phys. Rev. 168, 550, 563 (1068).
14. P. Jung and W. Triftshäuser, Phys. Rev. 175, 512 (1968).
15. A. H. Snell in α-, β-, AND γ-RAY SPECTROSCOPY, E. K. Siegbahn, Amsterdam: North Holland Publ. Co., 1965, p. 1545.
16. U. Gonser and H. Wiedersich, J. Phys. Soc. Japan 18, Suppl. II, 47 (1963).
17. P. Hannaford, C. J. Howard, and J. W. G. Wignall, Phys. Letters 19, 257 (1965).
18. P. Hannaford and J. W. G. Wignall, phys. stat. sol. 35, 809 (1969).
19. S. I. Bondarevskii and P. P. Seregin, Sov. Phys.-Sol. State 10, 2736 (1969), a translation of Fizika Tverdogo Tela 10, 3454 (1968).
20. A. N. Murin, B. G. Lure, S. I. Bondarevskii, and P. P. Seregin, Sov. Phys.-Sol. State 10, 2207 (1969), a translation of Fiz. Tver. Tela 10, 2803 (1968).
21. A. N. Murin, S. I. Bondarevskii, and P. P. Seregin, Sov. Phys.-Sol. State 12, 857 (1970), a translation of Fiz. Tver. Tela 12, 1095 (1970).
22. W. G. Berger, J. Fink, and F. E. Obenshain, Phys. Letters 25A, 466 (1967).
23. R. Shnidman, Y. K. Lee, and J. C. Walker, Report of AEC Contract AT(30-1)-2028, Sept. 1968, p. 26.
24. R. L. Lambe and D. Schroeer, Bull. Am. Phys. Soc. 16, 24 (1971).

25. E. P. Stepanov and A. Yu. Aleksandrov, JETP Letters $\underline{5}$, 83 (1967), a translation of ZhETF Pis'ma $\underline{5}$, 101 (1967).
26. W. G. Berger, Z. Physik $\underline{225}$, 139 (1969).
27. N. Hershkowitz, S. A. Wender, and L. W. Oberley, to be published.
28. C. C. Jacobs, Jr., and N. Hershkowitz, Phys. Rev. $\underline{B1}$, 839 (1970).
29. F. de S. Barros, D. Hafemeister, and J. P. Vincent, J. Chem. Phys. $\underline{52}$, 2865 (1970).
30. K. Matsui, R. R. Hasiguti, and H. Onodera, to be published in the Proceedings of the International Conference on Radiation Effects in Semiconductors, Albany, New York, U.S.A. (August, 1970).
31. D. Seyboth, F. E. Obenshain, and G. Czjzek, Phys. Rev. Letters $\underline{14}$, 954 (1965).
32. G. Czjzek, J. L. C. Ford, Jr., F. E. Obenshain, and D. Seyboth, Phys. Letters $\underline{19}$, 673 (1966); and G. Czjzek, J. L. C. Ford, Jr., J. C. Love, F. E. Obenshain, and H. H. F. Wegener, Phys. Rev. Letters $\underline{18}$, 529 (1967).
33. E. T. Ritter, P. W. Keaton, Jr., Y. K. Lee, R. R. Stevens, Jr., and J. C. Walker, Phys. Rev. $\underline{154}$, 287 (1967).
34. K. A. Hardy, D. C. Russell, and R. M. Wilenzick, Phys. Letters $\underline{27A}$, 422 (1968).
35. J. Fink and P. Kienle, Phys. Letters $\underline{17}$, 326 (1965).
36. J. Fink, Z. Physik $\underline{207}$, 225 (1967).
37. W. Mansel, G. Vogl, H. Wenzl, and D. Barb, phys. stat. sol. $\underline{40}$, 461 (1970).
38. J. B. Gibson, A. N. Goland, M. Milgram, and G. H. Vineyard, Phys. Rev. $\underline{120}$, 1229 (1960).
39. C. Erginsoy, G. H. Vineyard, and A. Englert, Phys. Rev. $\underline{133}$, A595 (1964).
40. C. Erginsoy, G. H. Vineyard, and A. Shimizu, Phys. Rev. $\underline{139}$, A118, (1965).
41. P. H. Dederichs, CHR. Lehmann, and H. Wegener, phys. stat. sol. $\underline{8}$, 213 (1965).
42. F. Seitz and J. S. Koehler in SOLID STATE PHYSICS, Vol. 2, edited by F. Seitz and D. Turnbull, N. Y. Academic Press, 1956, p. 305.
43. J. G. Mullen, Phys. Letters $\underline{15}$, 15 (1965).
44. B. H. Zimmermann, H. Jena, G. Ischenko, H. Kilian, and D. Seyboth, phys. stat. sol. $\underline{27}$, 639 (1968).
45. I. I. Lukaschevich, V. V. Sklarevskii, K. P. Aleshin, B. N. Samoilov, E. P. Stepanov, and N. I. Filippov, JETP Letters $\underline{3}$, 50 (1966), a translation of ZhETF Pis'ma $\underline{3}$, 81 (1966).

46. R. L. Cohen and H. J. Guggenheim, Nucl. Instr. and Meth. 71, 27 (1969).
47. B. Khurgin, S. Ofer, and M. Rakavy, Phys. Letters 33A, 219 (1970).
48. J. Christiansen, E. Recknagel, and G. Weyer, Phys. Letters 20, 46 (1966).

LOCAL MAGNETIC MOMENTS AND THE MOSSBAUER EFFECT

Brian B. Schwartz* and R.B. Frankel

Francis Bitter National Magnet Laboratory[+]
Massachusetts Institute of Technology
Cambridge, Massachusetts 02139

INTRODUCTION

In this brief paper we will describe the use of the Mossbauer effect to study the behavior of the magnetic properties of impurities in metals. This problem has been named the "localized moment problem" and in recent years has been studied quite thoroughly both theoretically and experimentally.[1] At the beginning of the last decade, it was believed that significant theoretical and experimental progress had been made. Based on earlier concepts of Friedel,[2] both Anderson[3] and Wolff[4] presented a sound theoretical model for studying the magnetic behavior of a single impurity in a host metal. At about the same time systematic studies of low concentration impurities in alloys of the 4d transition metals had revealed rules for the occurrence and definition of a magnetic impurity.[5] At this point in the study of the localized moment problem most researchers felt that significant progress had been made and ultimately a first principle understanding of the occurrence of elemental magnetism would soon be presented. This optimism was based on the historical belief that after solving the single impurity problem one could go on to the two impurity problem and then the many impurity problem. Ultimately it was felt that a magnetic metal like Fe could be considered as consisting of magnetic impurities at each lattice site. Unfortunately even the single impurity problem has turned out to be more difficult than first realized and a multitude of magnetic behaviors has been associated within even the single impur-

21

ity limit depending upon the nature of both the impurity
and the host.

Several experimental techniques can be used to study
the behavior of a localized magnetic moment at the impurity
site itself and with the surrounding conduction electrons,
nuclei and other impurities. Both local and long range
effects can be studied by microscopic and macroscopic
measurements. All the interpretations are based on the fact
that both the core electron spin polarization at the impurity site and the extended conduction electron spin polarization around the impurity are proportional to the magnetization of the localized moment associated with the impurity.
A listing of both the microscopic and macroscopic experimental techniques for observing the local and long range
effects of magnetic impurities has been given in detail by
Jaccarino.[6]

The hyperfine interaction at the impurity nucleus
allows the study of the localized moment itself. For
transition metal impurities such as Fe, there are several
contributions to the hyperfine field all of which are
proportional to the impurity magnetization $\langle S_z \rangle$. If the
impurity behaves like an isolated magnetic moment, then the
magnetic field and temperature dependence of the hyperfine
field should simply be a Brillouin function

$$H_{hyp} = H_{sat} \, B_S \left(\frac{g u_B H_o}{k_B T} \right)$$

where H_{sat} is the saturation hyperfine field for $B_S = 1$ and
is proportional to the total impurity moment $\langle S \rangle$ and
includes contributions from the polarized s-conduction band
via the Fermi contact term, H_o is the applied external field
and T the temperature. The Brillouin behavior is correct
as long as the impurity is in thermal equilibrium in a time
which is short compared to the Larmor precession time. If,
however, the electronic relaxation time becomes longer than
the Larmor precession time (e.g., at very low T), then the
total hyperfine field no longer follows a Brillouin function
although H_{sat} is still proportional to the magnitude of the
impurity moment. Although some systems seem to show a
Brillouin behavior for the magnetization of the local impurity, under more careful experimentation very few systems are
ideal. The susceptibility, the NMR linewidth and the impurity hyperfine field usually display anomalous low temperature

behavior. This is especially true for impurities which do
not display a Curie susceptibility at low temperature and in
alloys with anomalous resistive behavior such as a resistance
minimum.

   In a recent paper[7] a short review of the localized
moment problem was presented and in this paper we will
concentrate on describing some of the recent theoretical and
experimental developments. In particular we will discuss
the use of the Mossbauer effect to study the microscopic
effects associated with the Kondo effect on the magnetic
behavior of Fe in Cu at low temperature. The FeCu alloy
system is now considered the classical Kondo system and is
the one studied most extensively.[1] We will contrast the
behavior of the Mossbauer spectrum of Fe in Rh with the one
of Fe in Cu. FeRh alloys unlike FeCu alloys do not show
the usual resistance minimum associated with the Kondo effect
effect, however, its magnetic susceptibility as well as its
Mossbauer hyperfine field indicate an unusual magnetic
moment behavior associated with the Fe impurities. The last
part of this paper will deal with recent experimental
attempts to distinguish between two models for the occurrence
of a local magnetic moment. The continuous appearance of a
localized moment is usually associated with the Friedel-
Anderson-Wolff picture whereas the discontinuous appearance
of a moment depending upon the local environment of the
impurity is usually associated with a model due to Jaccarino
and Walker.[8] The microscopic Mossbauer probe is an excellent
tool for studying and differentiating these two models.
Studies on Fe in $Nb_{1-x}Mo_x$ alloys, in which a moment seems to
appear when $x = 0.4$, indicate the Jaccarino-Walker picture to
be appropriate however, even in this case additional compli-
cations associated with the Kondo effect seem to be present.

                       CLASSICAL BEHAVIOR

   In the simplest theory, the localized moment problem
can be separated into two parts: 1) If a localized moment
exists in a metal, what is its behavior right at the impurity site and its immediate environment? 2) What conditions
both for the impurity and the host metal are necessary in
order to have a localized moment form? These two questions
presume that the formation of the localized moment and its
interaction with the surroundings can be separated. However,
in more realistic models, the conditions for the formation

of the moment and its interaction with the host are linked in a quite complicated feedback system of equations.

The classical answer to question 1 can be obtained using s-d Hamiltonian proposed by Zener[9] to study the magnetism of the transition metals. Zener proposed a model in which d electrons were assumed to be localized and the s conduction electrons itinerant. The interaction can be written in the form

$$H_{s-d} = - \sum_i J(r - R_i) \, \sigma(r) \cdot S_i$$

where $J(r - R_i)$ is the s-d exchange coupling integral, $\sigma(r)$ is the spin density of the conduction electrons and S_i is the localized spin at the ith lattice sites.

An early attempt to quantify microscopically the conditions under which a local moment would form on an impurity (question 2) was presented by Friedel[2] and later developed by Anderson[3] and Wolff.[4] These models all include the following terms: 1) the kinetic energy of the conduction electrons; 2) the local energy E_d of the impurity state, 3) the s-d admixture element V_{sd} which acts to broaden the local-ized state leading to a width $\Delta = \pi N_c(0) V_{sd}^2$ where $N_c(0)$ is the host metal density of states at the Fermi surface. Within the Hartree-Fock approximation the local susceptibil-ity of the impurity using the Anderson Hamiltonian is

$$\chi_{imp} = \frac{2 \mu_B^2 N_d(0)}{1 - N_d(0)U}$$

where $N_d(0)$ is the density of states of the impurity d level at the Fermi surface and for one electron per impurity is given by $N_d(0) = 1/(\pi\Delta)$. Thus the criteria for the forma-tion of a local moment is the divergence of χ_{imp} leading to the condition $1 - N_d(0)U \leq 0$ which is equivalent to $[U/(\pi\Delta)] > 1$. This simple formula was used to interpret the classical susceptibility measurements by Clogston et al.[8] on the magnetic behavior of 1% Fe in binary alloys of neigh-boring 4d transition metals. In a rigid band model for the density of states of the 4d band electrons, the impurity width and thus $\Delta = V_{sd}^2 N_{4d}(0)$ varies dramatically while U remains virtually the same as one changes the d band occu-pation from 0 to 10 electrons. When $N_d(0)$ is large near the ends of the 4d band, Δ is narrow and $N_d(0)U$ is expected to

be large satisfying the condition for the divergence of the
local moment susceptibility. Thus one observes the appearance and disappearance of a localized moment with alloying
depending upon the value of $N_d(0)U$.

Higher order many body effects have been shown to
suppress the divergence of the susceptibility and thus a
more rigorous treatment can be used even for a single impurity. The correlations can be included in the calculation of
the dynamic frequency dependent susceptibility $\chi(\omega)$ in
which the spin moment decays with a characteristic lifetime
τ_{sf} called the spin fluctuation lifetime. τ_{sf} increases
with U but never becomes infinite even when $(U/\pi\Delta) > 1$.
Thus in this more rigorous theory the observation of a spin
moment will depend upon whether the fluctuations of the spin
are sufficiently slow compared with the time scale of the
experimental probe.

ANOMALOUS IMPURITY MOMENT BEHAVIOR

The simplicity of the Friedel-Anderson-Wolff model plus
its physical appeal and the experiments by Clogston et al.
gave hope that the single impurity problem was relatively
solved. In 1964, however Kondo[10] showed that the resistance minimum seen in such alloys as Fe in Cu could be interpreted as being due to a higher order scattering between the
magnetic Fe impurity and the Cu conduction electrons. Kondo
showed that this higher order scattering increased the resistivity as log T at low temperature and was the result of a
strong correlation between the magnetic impurity and the
conduction electron. Kondo started with the Zener s-d Hamiltonian and showed that the resistance due to magnetic impurities to 2nd order in the Born approximation is given by

$$\rho_T(T) = AT^5 + c\rho_m \left[1 - N(0)J \ln(T/D) \right]$$

where AT^5 is the temperature dependence of the resistance
associated with phonon scattering, c the concentration of
impurities, ρ_m the 1st Born approximation scattering resistance per impurity, J the s-d coupling constant, and D a
measure of the width of the conduction band. $\rho_T(T)$ shows a
minimum which is weakly dependent upon concentration,
$T_{min} \propto c^{1/5}$. The expansion must break down when the high
order term $-N(0)J \ln(T/D) \simeq 1$ giving rise to the Kondo

temperature $T_K = D_{exp}[-1/N(0)J]$. Much below this characteristic temperature very strong correlations exist between the d spin electrons localized on the impurity site and the conduction electrons. These spin correlations considerably modify the low temperature spin behavior both on the impurity site and on the host metal electrons and ions in the neighborhood of the impurity. In other words, the arbitrary separation of the problem into two parts posed above breaks down. The Mossbauer effect being a microscopic probe is an ideal tool to study the formation and effects of these low temperature correlations. In what follows we will contrast the susceptibility, resistivity and Mossbauer measurements for FeCu with that of FeRh.

Besides the resistivity, the Kondo effect is believed to affect the magnetization of the impurity which can be seen in the measurement of the susceptibility. Below the Kondo temperature low temperature resistivity measurements indicate that the log T divergence levels off approaching a saturation value corresponding to the unitarity limit of scattering per impurity. The unitarity limit is expected when the scattering potential becomes so strong that it actually binds the electrons. Although the theory and experiments have advanced considerably and are quite sophisticated it is easiest to appreciate some of the physical difficulties by relying first on some earlier simpler interpretations. The unitarity limit in the resistivity seems to be the result of a complicated singlet spin coupling between the conduction electron spin and the impurity spin. This coupling of the impurity and conduction electron spins leads to an effective cancellation of the moment associated with the impurity. Thus at low temperature the susceptibility no longer follows a Curie law but seems to vary as
$\chi(T) \propto C/T + T_K$ and no longer diverges as $T \to 0$. Examples of the low temperature resistivity and susceptibility behaviors of Fe in Cu are given in Figs. 1a and 2a. Thus, it seems that associated with the Kondo effect is a characteristic energy $k_B T_K$ which represents the effective correlation energy between the impurity and the conduction electron spin. In addition because of the singlet nature of the coupling, the effective spin moment of the impurity seems quenched and equal to zero at low temperature. First, this quenching of the spin moment should be observable using the microscopic Mossbauer probe at the impurity site and socond, this quenching of the impurity moment should be capable of being destroyed by the application of a magnetic field such that

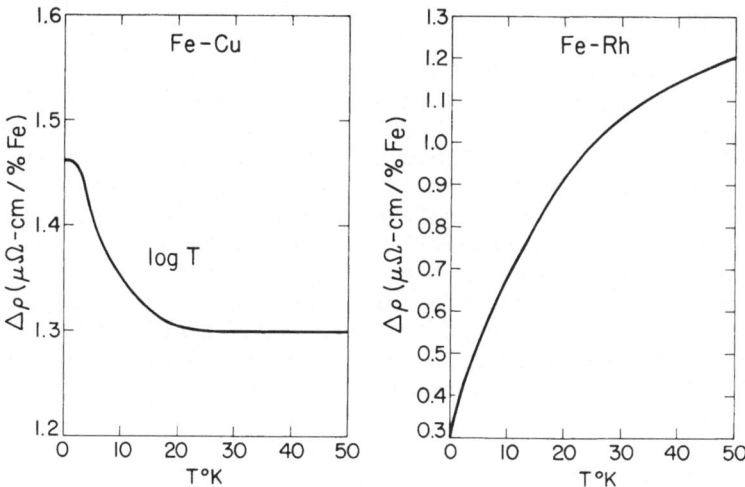

Fig. 1.   The anomalous resistivity per impurity of Fe in Cu and the total resistivity of 1% Fe in Rh.

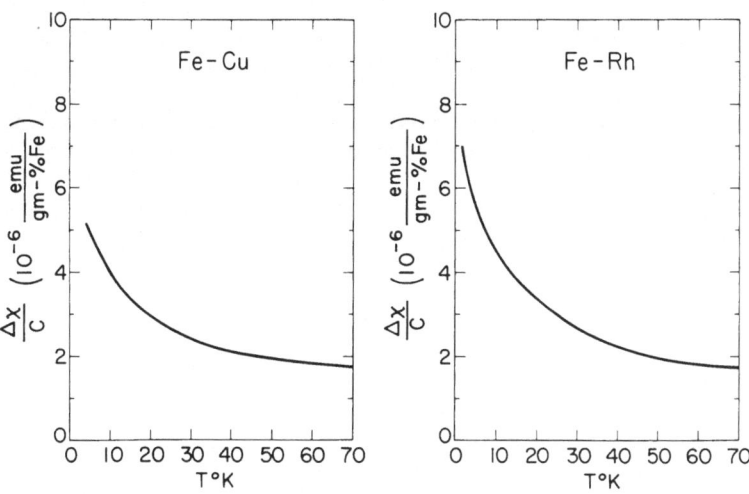

Fig. 2.   The low temperature susceptibility per impurity of Fe in Cu and Fe in Rh.

$\mu_B H_0 \approx k_B T_K$.

The Kondo temperature of Fe in Cu is believed to be on the order of 10 K which is ideal experimentally since it can conveniently be studied at temperature T both above and below T_K and a magnetic field of about 100 kG is expected to have a strong destructive effect with respect to the formation of the spin correlated state. The hyperfine field at the Fe nucleus is primarily due to the core-polarization produced by the moment on the Fe and the conduction electron spin polarization associated with the moment. Thus any changes in the behavior of the measured hyperfine field reflect the local nature of the correlations associated with the formation of the low temperature Kondo spin compensated state.

In Fig. 3 we show the hyperfine field measurements of Fe^{57} in Cu as a function H_0/T.[11] The three solid line curves represent the observed hyperfine field at constant magnetic field and decreasing temperature. Note the hyperfine field does not follow a Brillouin function and that the saturation hyperfine field as $T \to 0$ is not a constant but depends upon the applied field. The growth observed in H_{sat} with H_0 can be interpreted as the breaking up of the spin compensated state and the growth of the moment on the iron impurity. From a plot of $H_{sat}(H)$ vs H_0 as given in Fig. 4a an order of magnitude value for T_K can be obtained. The high temperature and therefore Brillouin behavior of the hyperfine field can be used to obtain H_{sat} which represents the saturation hyperfine field which would have been obtained for $H_0/T \to \infty$ had not the spin compensated state formed at low temperature. The value for $\mu_B H_K$ gives a value for $k_B T_K$ on the order of the Kondo energy. The actual approach to saturation of the moment as a function of H_0 for $T < T_K$ has been studied theoretically by Nam and Woo[12] and by Ishii.[13] They find that the moment starts off increasing linearly with field and eventually curves over and saturates at high field. This region where $\mu H \gg kT_K$ has recently been investigated by Maley and Taylor[14] in Fe in Mo which has a $T_K \sim 0.25$ K. In Fe in Cu for low applied field H_0, $H_{sat}(0)/H_{sat} \ll 1$ confirming in part the singlet nature of the spin compensated state for $T \ll T_K$. More details of the experiment are given by Frankel et al.[11] and a review of Mossbauer and NMR as well as other studies on the Kondo effect are given in the review paper by Heeger.[1]

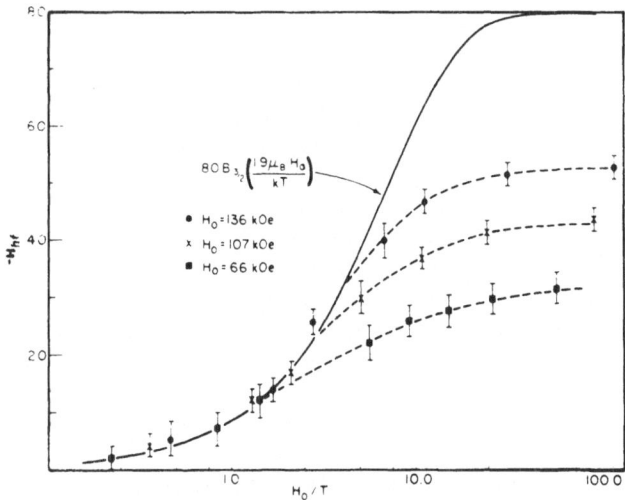

Fig. 3. The hyperfine field for Fe^{57} in Cu as a function of H_0/T for three values of the external field.

In addition to Fe in Cu the low temperature high field Mossbauer measurements have been performed in other alloys including Fe in Rh.[15,16,17] Fe in Rh is a very interesting dilute alloy system. In Figs. 1b and 2b we plot the resistivity and susceptibility of Fe in Rh and contrast its behavior with that of Fe in Cu. The resistance decreases with temperature and does not show any resistance minimum, however, the susceptibility of Fe in Rh except for a scaling factor looks very much like that of Fe in Cu. One can therefore ask, since the magnetic susceptibility of FeRh is essentially similar to that of FeCu, what are the micro- scopic details of the magnetic behavior of the Fe impurity? If the localized magnetic behavior is the same, then why is the resistivity of FeCu and FeRh so different?

High field and low temperature Mossbauer measurements have been performed on Fe in Rh and the hyperfine field as a function of applied field shows a behavior similar to that of Fe in Cu. The saturation hyperfine field $H_{sat}(H_0)$ vs H_0 is shown in Fig. 4b and seems to grow linearly with applied field. The limiting saturation hyperfine field yields a Kondo field $H_K \approx 300$ kG and represents a compensation energy $\cup_B H_K \approx k_B T_K \approx 30$ K which agrees well with the compensation temperature obtained from the fit to susceptibility data.

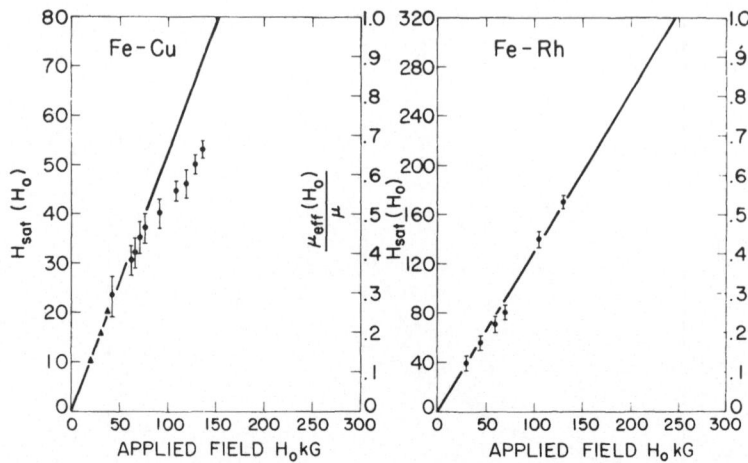

Fig. 4. The saturation value of the magnetic hyperfine field at low temperature as a function of applied magnetic field for Fe in Cu and Fe in Rh.

The interpretation of the growth of the low temperature moment in terms of the breaking up of the spin compensated state is among one of the models used to explain the Mossbauer data. Another approach is to use the spin fluctuation model in which the characteristic spin fluctuation energy $\hbar\omega_{sf} \approx k_B T_K$ such that for $T > T_K$ there appears to be a moment and for $T < T_K$ there is no moment because the spin moment fluctuates too fast leading to no net moment. There are various theories to explain the unusual resistivity behavior shown in Fig. 1b. In one model due to Knapp,[18] the Rh host is believed to be composed of both d electrons and s electrons. The d electrons partake in the spin compensation of the Fe impurity moment and the s conduction electrons carry the current and scatter off of the compensated state. Since the resistivity is proportional to the square of the Fe impurity moment, one expects the resistivity as a function of T to be proportional to the square of the effective temperature dependent moment or equivalent $\rho(T) \propto T \chi(T) \propto \mu_{eff}^2(T)$. This simple two-band model seems to give a good fit to the resistivity and susceptibility data. Another model making use of the spin fluctuation theory has recently been worked out in detail by Doniach.[19] Very low temperature resistivity measurements by Foner et al.[20]

seem to indicate some difficulty in obtaining a unique
characteristic spin fluctuation temperature. High field
magnetoresistance measurements by Foner[20] on FeCu and FeRh
also confirm the anomalous magnetic behavior of FeRh.

THE LOCAL ENVIRONMENT EFFECT

In the study of the formation of a local moment, macroscopic measurements are unable to determine the microscopic
behavior of the local moment uniquely. The interpretation
of the magnetic properties of Fe impurities in 4d transition
metal binary alloys assumed that when the Fe impurities are
magnetic they all have the same net moment. The rigid band
uniform model has been challenged by Jaccarino and Walker[8]
who proposed that the magnetic moment on an impurity occurs
discontinuously and its appearance depends upon the local
environment about the impurity rather than the Anderson
$U/(\pi\Delta) > 1$ requirement. The experimental confirmation of
this local environment model was observed in interpreting the
Co^{59} NMR spectrum in the $Rh_{1-x}Pd_x$ alloys. As x increased the
Co^{59} signal decreased in intensity but did not shift in
frequency from that observed for Co in pure Rh where there
is no moment. If all the Co atoms had the same average
moment, then as the concentration of Pd increased above that
necessary for magnetic behavior no dimunition of the signal
would have been seen, but rather a gradual shift in the
resonance due to the hyperfine field generated by the
localized moment formed on the Co.

Jaccarino and Walker were able to account for the
results by assuming that a given impurity was limited to
having either no moment or its full moment depending on
whether or not it has a minimum number of Pd near neighbors.
Then the intensity of the Co^{59} NMR resonance and the average
moment per impurity could be calculated from a binomial
like distribution.

The same behavior has been observed in Fe and Co-doped
$Nb_{1-x}Mo_x$. Neither Fe or Co are magnetic in Nb metal, but
both have moments in Mo. In the Co-doped alloys NMR in Co^{59}
by Brog et al.[21] detect two resonances which change in intensity as a function of x, one of which is due to those nuclei
which see only the external field and the other of which is
due to those nuclei whose resonance is shifted by a hyperfine interaction due to the presence of a localized moment

on the impurity.   The important point is that as x changes,
only the relative intensity of the two resonances, not the
positions of the resonances changes.

The Fe-doped alloys have been studied by Mossbauer
spectroscopy and show similar effects.[22,23,24]   In Fig. 5
we show the Mossbauer spectrum for Fe in pure Nb and pure
Mo.   In Nb there is no moment so that the low temperature
hyperfine field at the nucleus is the same as the applied
field 75 kG.   In Mo there is a moment and the hyperfine
field is -115 kG, which leads to a 40 kG separation of the
outer lines in the Mossbauer spectrum.   The Mossbauer experiments have been performed on binary alloys of $Nb_{1-x}Mo_x$ for
x = 0, 0.2, 0.4, 0.6, 0.8 and 1.0.   Above x = 0.4 the hyperfine field shows two distinct sites indicating the local
environment effect of full moment Fe impurity coexisting
with Fe impurities with no moment.   The temperature and field
dependence of the Mossbauer data is actually more complicated due to the appearance of Kondo like spin compensation

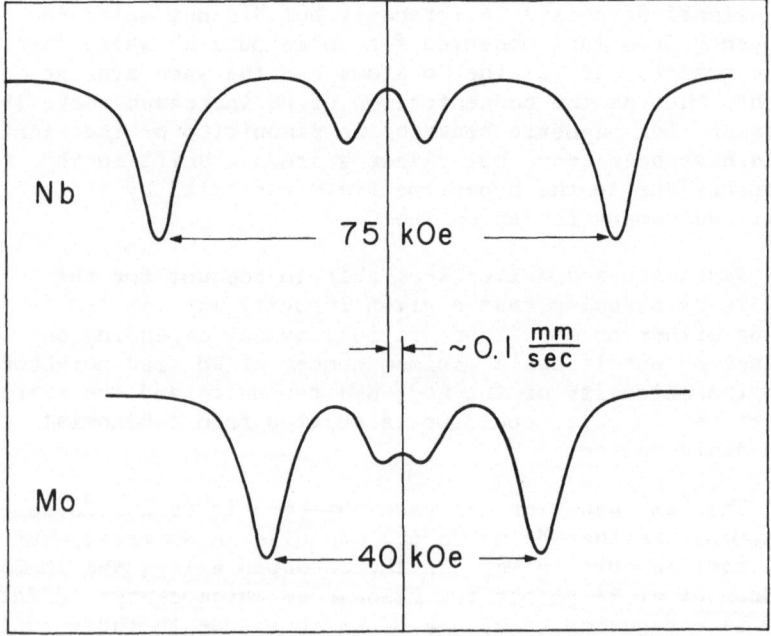

Fig. 5.   The Mossbauer spectrum at 4.2 K and 75 kG for Fe
in pure Nb and pure Mo.

effects at low T.

Recently, the relation between the Jaccarino-Walker effect and the Anderson model was studied by Kim.[25] Kim discusses the effect of the self energy of an impurity due to the interaction with other surrounding impurities. The real and imaginary parts of the self energy give rise respectively to the shift and broadening (or narrowing) of the impurity state. The local environment effects of Jaccarino and Walker originate from the fact that the self energy of an impurity, the broadening and the shifting of the impurity level, depends on the distribution of the other impurities in its immediate neighborhood.

CONCLUSION

In spite of the large amount of theoretical and experimental work done on the dilute alloy problem, it remains something of a mystery. The single impurity problem is an especially difficult many body problem as compared to superconductivity. For superconductivity the electron-phonon interaction so dominates the superconducting behavior of metals that normal metal parameters such as density of states and other electron-electron interaction can be neglected in a first order calculation. Essentially, a law of corresponding states exists between all the superconductors with the transition temperature or energy gap being a measure of the correlation effects. This is not true for a single impurity interacting with the host metal. There are many interactions which must be included to discuss the magnetic properties of the impurity and include the s-d overlap interaction, the d-d interaction, and even the s-s exchange interaction. In addition, details of the density of states of the host metal can effect the properties of the impurity state. Thus the very strongly correlated behavior involving V_{sd}, U, E_d and the density of states can lead to a large multitude of behaviors each of which may have little in common with one another. The simple Hartree-Fock treatment first proposed by Anderson becomes extremely complicated when higher order correlations are taken in account even for idealized parabolic host bands and a single orbital. Unfortunately, it may turn out that a general solution of the localized moment problem will not be possible, but subclasses of the general problem can be used to interpret the experimental results. And while we are waiting for the perfect theory,

there are many experimental studies that remain to be done
to present even a phenomenological understanding of the
problem.

REFERENCES

*Also Physics Department, Massachusetts Institute of
 Technology.
†Supported by the U.S. Air Force Office of Scientific
 Research.

1. J. Kondo in Solid State Physics, F. Seitz, D. Turnbull,
    and H. Ehrenreich, eds. (Academic Press, New York, (1969),
    Vol. 23, p. 184; A.J. Heeger in Solid State Physics,
    F. Seitz, D. Turnbull, and H. Ehrenreich, eds. (Academic
    Press, New York, 1969), Vol. 23, p. 284; M.D. Daybell
    and W.A. Steyert, Rev. Mod. Phys. 40, 380 (1968).

2. J. Friedel, Nuovo Cimento (Suppl.) 7, 287 (1958).

3. P.W. Anderson, Phys. Rev. 124, 41 (1961).

4. P.A. Wolff, Phys. Rev. 124, 1030 (1961).

5. A.M. Clogston, B.T. Matthias, M. Peter, H.J. Williams,
    E. Corenzwit, and R.C. Sherwood, Phys. Rev. 125, 541
    (1962).

6. V. Jaccarino, J. Appl. Phys. 39, 1166 (1968).

7. B.B. Schwartz in Hyperfine Interactions Detected by
    Nuclear Radiation (Israel Science Services Inc.) to
    be published.

8. V. Jaccarino and L.R. Walker, Phys. Rev. Letters 15,
    258 (1965).

9. C. Zener, Phys. Rev. 81, 440 (1951).

10. J. Kondo, Prog. Theor. Phys. (Japan) 32, 37 (1964).

11. R.B. Frankel, N.A. Blum, B.B. Schwartz, and D.J. Kim,
    Phys. Rev. Letters 18, 1050 (1967).

12. S.B. Nam and J.W.F. Woo, Phys. Rev. Letters 19, 649 (1967).

13. H. Ishii, Progr. Theor. Phys. (Japan) 40, 201 (1968).

14. M.P. Maley and R.D. Taylor, to be published.

15. N.A. Blum, J. Chappert, R.B. Frankel and B.B. Schwartz, Bull. Am. Phys. Soc. 11, 410 (1968).

16. T.A. Kitchens, W.A. Steyert, and R.D. Taylor, Phys. Rev. 138, A467 (1965).

17. T.A. Kitchens, P.P. Craig, and R.D. Taylor, in Mossbauer Effect Methodology, I.J. Gruverman, ed. (Plenum Press, New York, 1969), Vol. 5.

18. C. Knapp, Phys. Letters 25A, 114 (1967).

19. A.B. Kaiser and S. Doniach, Intern. J. Magnetism 1, 11 (1970).

20. S. Foner, private communication.

21. K.C. Brog and W.H. Jones, Jr., Phys. Rev. Letters 24, 58 (1970).

22. H. Nagasawa and N. Sakai, J. Phys. Soc. Japan 27, 1150 (1969).

23. N.A. Blum, R.B. Frankel, and R.P. Guertin, Bull. Am. Phys. Soc. 15, 262 (1970).

24. L.J. Swartzendruber, Phys. Rev. to be published.

25. D.J. Kim, Phys. Rev. B 1, 3725 (1970).

SPECTROSCOPY

Analytical Determination of Magnetic and Quadrupole
Hyperfine Parameters from Mössbauer Spectra

P.G.L. Williams and G.M. Bancroft

Department of Chemistry
The University of Western Ontario
London 72, Canada

A general analytical method is outlined for
determining the magnetic and quadrupole hyperfine parameters
directly from an observed Mössbauer spectrum for the
majority of the common nuclear spins.

The Hamiltonian matrix describing the interaction of
the field system with the nucleus is set up for nuclear
spins up to $I = \frac{9}{2}$. From the above matrix, a secular
determinant is constructed which is expanded to give the
secular polynomial, whose $2I + 1$ roots are the energy
eigenvalues of the nucleus. These eigenvalues may be
determined easily from many observed spectra. By equating
the sums of the products of these roots to the coefficients
of the powers of E in the secular polynomial, equations are
obtained which are solved to give the hyperfine parameters.

To illustrate the usefulness of the above method, the
hyperfine parameters have been obtained for iodine and
neptunium Mössbauer spectra reported in the literature.
The significant advantages and possible difficulties in the
method are critically examined.

INTRODUCTION

A Mössbauer spectrum consists of a number of lines
arising from transitions between the sub-levels of the
ground and first excited states of the absorber nucleus.

39

A nucleus of spin I has a degeneracy of $2I + 1$ in a field-free environment (free space). The degeneracy of these sub-levels may be removed by the interaction of the nucleus with any local magnetic and electric fields, and thus the positions of the lines in the Mössbauer spectrum arising from the permitted transitions between these sub-levels are determined by the form and magnitude of this interaction.

If the fields, and the appropriate nuclear moments are known, it is possible to calculate the energy level diagram, and thus the observed spectrum. This is done by writing down the Hamiltonian describing the interaction between the nucleus and its environment, and setting up the Hamiltonian matrix $\langle m_i | \mathcal{H} | m_j \rangle$ over the spin component basis set $| m_i \rangle$. Diagonalisation of this matrix gives the eigenvalues of the energy of the nuclear sub-levels - that is, the energy level diagram for the particular nuclear state. If this calculation is carried out for both the ground and excited states in order to give the complete energy level diagram, the line positions in the Mössbauer spectrum can be determined directly (with respect to an arbitrary centre shift).

However, the problem is almost invariably the reverse of this, in that it is desired to calculate the hyperfine parameters from an observed spectrum. Numerical [1, 2] and graphical [3, 4] methods have been developed for solving this problem, but these have certain disadvantages. Thus the graphical method, which involves plotting the energy level diagram as a function of certain chosen hyperfine parameters, fitting the observed spectrum to this diagram and reading off the value of the particular parameter, is not very accurate, while numerical methods based on an iterative fitting process require the use of a complex fitting program and reasonable initial estimates of the parameters. In this article, a general analytical method is described whereby the hyperfine parameters may in many cases be calculated directly from the observed spectrum by a relatively simple analytical method.

AN INITIAL OUTLINE OF THE METHOD

The Hamiltonian describing the quadrupole and/or magnetic interaction may be written down for the particular nuclear state under consideration, and the Hamiltonian matrix over the basis set of the spin components thus constructed. This leads to a secular determinant which may be expanded to give a polynomial in E, the energy, whose n roots (where n = 2I + 1) are the eigenvalues of the energy of the nucleus in the environment described by the Hamiltonian. This polynomial may be expressed as follows:

$$E^n + a_{n-1}E^{n-1} + a_{n-2}E^{n-2}\ldots+ a_iE^i\ldots+a_1E + a_o = 0 \qquad (1)$$

where the coefficients a_i are analytical functions of some or all of the hyperfine parameters.

Now if the roots of this (i.e. the energy levels of the nuclear state) are E_i for i = 1 to n, then

$$(E - E_n)(E - E_{n-1})\ldots(E - E_i)\ldots(E - E_1) = 0$$

or $E^n - E^{n-1}(\sum_i E_i) + E^{n-2}(\sum_{i>j} E_i E_j)\ldots+ (-1)^n E_1 E_2 \ldots E_n = 0 \quad (2)$

is identical to (1). Therefore, equating coefficients, we obtain:

$$a_{n-1} = -\sum_i E_i; \quad a_{n-2} = \sum_{i>j} E_i E_j; \ldots; \quad a_o = (-1)^n E_1 E_2 \ldots E_n \qquad (3.1 \rightarrow 3.n)$$

Equations (3) comprise a set of n equations relating functions of the hyperfine parameters (the coefficients a_i) to products of the energy levels of the nucleus. The energy levels of the nuclear state may be obtained from the energy level diagram, which may in turn be derived directly from the observed spectrum as is illustrated later. In many cases, these equations (3) may be solved directly to yield the hyperfine parameters.

In this article, the above equations are derived explicitly for nuclear spin states I = 1, $\frac{3}{2}$, 2, $\frac{5}{2}$, $\frac{7}{2}$ and $\frac{9}{2}$ for a pure quadrupole interaction, and for nuclear spin states I = 1, $\frac{3}{2}$, 2 and $\frac{5}{2}$ for a combined general magnetic and quadrupole interaction, and the application of this method to the treatment of two specific cases is given.

ANALYTICAL DETERMINATION OF THE QUADRUPOLE
HYPERFINE PARAMETERS

Half Integral Spins

In the absence of a magnetic hyperfine interaction, and if the nucleus concerned possesses a quadrupole moment, eQ, (i.e. if the nuclear spin ≥ 1), then the quadrupole hyperfine interaction is the only interaction removing the degeneracy of the nuclear spin components, and transitions between two such levels lead to a pure quadrupole Mössbauer spectrum.

The parameters which can be obtained for such a spectrum are e^2qQ, η, Qex/Qgr and δ, the centre shift [5]. The Hamiltonian representing the interaction of a nucleus, having spin I and quadrupole moment eQ, with an electric field gradient specified in the usual way by eq and η may be written [6]:

$$\mathcal{K} = \frac{e^2qQ}{4I(2I-1)}[3I_z^2 - I(I + 1) + \frac{\eta}{2}(I_+^2 + I_-^2)] \qquad (4)$$

where I is the total nuclear spin operator, $I_\pm = I_x \pm iI_y$, and I_x, I_y and I_z are the nuclear spin component operators.

From this, a matrix $\mathcal{K}$ may be constructed over the (2I +1)-fold basis set $|m_i\rangle$ in which $\mathcal{K}_{ij} = \langle m_i|\mathcal{K}|mj\rangle$. The reduction of $\mathcal{K}$ to diagonal form gives the 2I + 1 eigenvalues of the energy of the nucleus in the field system specified by eq and η.

Equivalently, a secular determinant

$$\left|\mathcal{K} - EI\right|_{ij} = \mathcal{K}_{ij} - E\delta_{ij}$$

where I is the unit matrix, may be constructed, and on expansion this yields a secular polynomial in E whose 2I + 1 roots are the eigenvalues of the energy of the nucleus in the field system eq, η.

The secular determinants for nuclei with spin $\frac{3}{2}$, $\frac{5}{2}$, $\frac{7}{2}$ and $\frac{9}{2}$ are evaluated, and that for spin $\frac{5}{2}$ is given as an example in the appendix. The determinants for $I = \frac{3}{2}$, $\frac{7}{2}$ and $\frac{9}{2}$ have a similar form. The following points about these determinants may be noted:

1. Writing the Hamiltonian as

$$\mathcal{K} = \mu(3I_z^2 - I(I + 1)) + \frac{\mu\eta}{2}(I_+^2 + I_-^2) \text{ where } \mu = \frac{e^2qQ}{4I(2I - 1)}$$

the elements on the leading diagonal arise from the first term only; that is, if $\eta = 0$, the eigenvalues of the energy are given by the diagonal elements, and thus m_I is a good quantum number. The off-diagonal elements arise from the second term in the Hamiltonian, and thus mixing between the spin components arises when $\eta \neq 0$. For $\eta \neq 0$, the m_I are not good quantum numbers with which to label the eigenstates of the Hamiltonian.

2. The determinants are symmetric across both diagonals, and consequently the secular polynomials obtained by expansion are perfect squares, indicating that the roots occur in pairs - i.e. Kramer's doublets.

The secular polynomials obtained by expansion of the above determinants are as follows:

$$I = \tfrac{3}{2}: [E^2 - 3\mu_{\frac{3}{2}}^2(3 + \eta^2)]^2 = 0 \tag{5}$$

$$I = \tfrac{5}{2}: [E^3 - 28\mu_{\frac{5}{2}}^2(3 + \eta^2)E - 160\mu_{\frac{5}{2}}^3(1 - \eta^2)]^2 = 0 \tag{6}$$

$$I = \tfrac{7}{2}: [E^4 - 126\mu_{\frac{7}{2}}^2(3 + \eta^2)E^2 - 1728\mu_{\frac{7}{2}}^3(1 - \eta^2)E$$
$$+ 945\mu_{\frac{7}{2}}^4(3 + \eta^2)^2]^2 = 0 \tag{7}$$

$$I = \tfrac{9}{2}: [E^5 - 396\mu_{\frac{9}{2}}^2(3 + \eta^2)E^3 - 950\mu_{\frac{9}{2}}^3(1 - \eta^2)E^2$$
$$+ 19008\mu_{\frac{9}{2}}^4(3 + \eta^2)^2E + 373248\mu_{\frac{9}{2}}^5(3 + \eta^2)(1 - \eta^2)]^2 = 0 \tag{8}$$

The above are equivalent to those given by Cohen [7]. The roots of the secular polynomials for each of the nuclear spin states (E_i) can be derived from the Mössbauer spectrum (see later). Thus the terms on the right hand sides of equations 3.1 to 3.n are known, and the terms on the left hand sides have been evaluated in the secular polynomials (5) to (8) above, in terms of μ and η. However, one of these equations ($a_{n-1} = \Sigma E_i$) cannot be used. In each of the secular polynomials (5) to (8) it is seen that the coefficient of E^{n-1} is zero. This is because the inter-action of the nucleus with an electric field gradient partially lifts the $(2I + 1)$-fold degeneracy of the nucleus, but does not displace the centre of gravity of the energy levels. Consequently $(2I - 1)/2$ equations are available to determine the two unknown parameters. Thus for $I = \tfrac{3}{2}$, μ and η cannot be separated. For $I = \tfrac{5}{2}$, two equations are available, and μ and η may be separated, while for $I > \frac{5}{2}$, there are redundant equations.

For $I = \frac{3}{2}$, equation (5) gives directly the standard result, i.e.:

$$E = \pm \frac{e^2 qQ}{4} \sqrt{\frac{3 + \eta^2}{3}} \qquad (9)$$

For $I = \frac{5}{2}$, a comparison of equation (6) with

$$E^3 - E^2 \sum_i E_i + E_i \sum_{i>j} E_i E_j - E_1 E_2 E_3 = 0$$

gives

$$-28\mu_{\frac{5}{2}}^2 (3 + \eta^2) = \sum_{i>j} E_i E_j \quad \text{and} \quad 160 \mu_{\frac{5}{2}}^3 (1 - \eta^2) = E_1 E_2 E_3$$

$$(10) \text{ and } (11)$$

Putting $C_1 = \dfrac{\overset{-\sum}{i>j} E_i E_j}{28}$ and $C_2 = \dfrac{E_1 E_2 E_3}{160}$

gives $C_1 = \mu_{\frac{5}{2}}^2 (3 + \eta^2)$ and $C_2 = \mu_{\frac{5}{2}}^3 (1 - \eta^2)$ (12) and (13)

Elimination of η gives $4\mu_{\frac{5}{2}}^3 - C_1 \mu_{\frac{5}{2}} - C_2 = 0$ (14)

This cubic equation may be solved for μ by the standard method [8], and substitution in equation (12) or (13) above gives η.

For $I = \frac{7}{2}$, equating coefficients in (7) with those of the appropriate polynomial gives the three equations:

$$-126\mu_{\frac{7}{2}}^2 (3 + \eta^2) = \sum_{i>j} E_i E_j \qquad (15)$$

$$1728\mu_{\frac{7}{2}}^3 (1 - \eta^2) = \sum_{i>j>k} E_i E_j E_k \qquad (16)$$

$$945\mu_{\frac{7}{2}}^4 (3 + \eta^2)^2 = E_1 E_2 E_3 E_4 \qquad (17)$$

Putting $C_3 = \dfrac{\overset{-\sum}{i>j} E_i E_j}{126}$ and $C_4 = \dfrac{\sum_{i>j>k} E_i E_j E_k}{1728}$

gives $\mu_{\frac{7}{2}}^2 (3 + \eta^2) = C_3$ and $\mu_{\frac{7}{2}}^3 (1 - \eta^2) = C_4$ (18) and (19)

which may be solved for μ and η as previously shown. In addition, there exists a redundant equation (17).

For $I = \frac{9}{2}$, comparing equation (8) with the appropriate polynomial and equating coefficients gives:

$$396\mu_{\frac{9}{2}}^{2}(3 + \eta^{2}) = -\sum_{i>j}E_{i}E_{j} \tag{20}$$

$$9504\mu_{\frac{9}{2}}^{3}(1 - \eta^{2}) = \sum_{i>j>k}E_{i}E_{j}E_{k} \tag{21}$$

$$19008\mu_{\frac{9}{2}}^{4}(3 + \eta^{2})^{2} = \sum_{i>j>k>1}E_{i}E_{j}E_{k}E_{1} \tag{22}$$

$$373248\mu_{\frac{9}{2}}^{5}(3 + \eta^{2})(1 - \eta^{2}) = -E_{1}E_{2}E_{3}E_{4}E_{5} \tag{23}$$

Any two of these four equations taken together will yield μ and η, and there are in addition two redundant equations.

Integral Spins

The secular determinants for $I = 1$ and $I = 2$ are derived. The determinants are no longer centrosymmetric, and $2I$ equations can be written down in terms of μ and η. For $\eta \neq 0$, the degeneracy of the $\pm m_{i}$ substates is removed, while the coefficient of E^{n-1} is again zero.

For $I = 1$, the coefficients in the polynomial

$$E^{3} + a_{2}E^{2} + a_{1}E + a_{o} = 0 \tag{24}$$

are: $a_{2} = 0$; $a_{1} = -\mu_{1}^{2}(3 + \eta^{2})$; $a_{o} = 2\mu_{1}^{3}(1 - \eta^{2})$ $\quad$ (25)

Comparison of (24) with:

$$E^{3} - E^{2}(\sum_{i}E_{i}) + E(\sum_{i>j}E_{i}E_{j}) - E_{1}E_{2}E_{3} = 0 \tag{26}$$

gives:

$$\mu_{1}^{2}(3 + \eta^{2}) = -\sum_{i>j}E_{i}E_{j} \quad (27) \text{ and } 2\mu_{1}^{3}(1 - \eta^{2}) = -E_{1}E_{2}E_{3} \quad (28)$$

Solution of the above two equations gives μ and η.

For $I = 2$, the coefficients in the polynomial

$$E^{5} + a_{4}E^{4} + a_{3}E^{3} + a_{2}E^{2} + a_{1}E + a_{o} = 0 \tag{29}$$

are: $a_{4} = 0$; $a_{3} = -21\mu_{2}^{2}(3 + \eta^{2})$; $a_{2} = -54\mu_{2}^{3}(1 - \eta^{2})$;
$a_{1} = 108\mu_{2}^{4}(3 + \eta^{2})^{2}$; $a_{o} = 648\mu_{2}^{5}(3 + \eta^{2})(1 - \eta^{2})$ $\quad$ (30)

Comparison of equation (28) with:

$$E^5 - E^4(\sum_i E_i) + E^3(\sum_{i>j} E_i E_j) - E^2(\sum_{i>j>k} E_i E_j E_k)$$

$$+ E(\sum_{i>j>k>\ell} E_i E_j E_k E_\ell) - E_1 E_2 E_3 E_4 E_5 = 0 \qquad (31)$$

yields four equations, any two of which can be solved for μ and η.

Determination of Energy Levels E_i

In order to use the equations derived in the last sections, the energy level diagram must be constructed and all the E_i values calculated. This is easily done if most of the lines are well resolved so that the line positions can be accurately obtained.

A Mössbauer nucleus with ground state spin $I = \frac{7}{2}$ and excited state spin $\frac{5}{2}$ is considered to illustrate the general procedure.

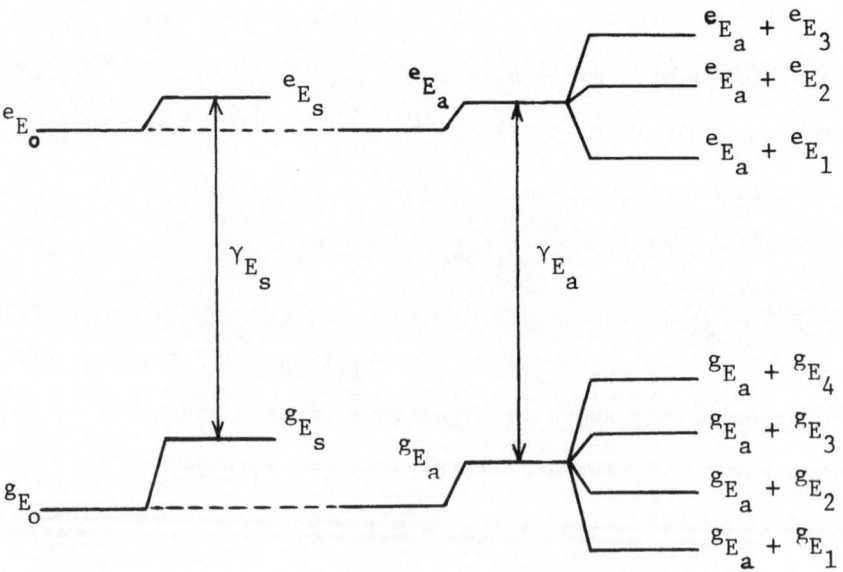

Free   Single-line   Free   Field-free General
space   source   space   absorber   absorber

It is seen that in a field-free absorber the single
Mössbauer line would lie at a velocity

$$\delta = {}^{\gamma}E_a - {}^{\gamma}E_s = ({}^{e}E_a - {}^{g}E_a) - ({}^{e}E_s - {}^{g}E_s)$$

where δ is the centre shift of the absorber with respect to
the particular source used. A general line would lie at
velocity L_k where

$$L_k = [({}^{e}E_a + {}^{e}E_j) - ({}^{g}E_a + {}^{g}E_i)] - ({}^{e}E_s - {}^{g}E_s)$$

for $i = 1 \rightarrow 4$ and $j = 1 \rightarrow 3$.

$$\therefore \ L_k = {}^{e}E_j - {}^{g}E_i + \delta \tag{32}$$

where $\Sigma {}^{e}E_j = \Sigma {}^{g}E_i = 0$

    Excited   Ground
     state   state

    For $\eta = 0$, the energy levels of the ground and excited
states may be labelled by the appropriate spin quantum
number m_I, since the Hamiltonian commutes with the spin
component operator I_z. The selection rule $\Delta m_I = 0, \pm 1$
limits the number of permitted transitions between the
energy levels of the ground and excited state. The ratio of
the intensities, as deduced from the Clebsch-Gordon coefficients serves to assign the lines, and the energy level
diagram can be constructed.

    For small values of η, in which the number of lines
capable of resolution is still the same as for $\eta = 0$, i.e.
mixing of states $m_I \pm 2$ into each state m_I by the second
term in the Hamiltonian is small, and thus m_I can still
serve as an approximate quantum number, the Clebsch-Gordon
coefficients may still be used as a guide to assignment.

    For large values of η, in which the eigenstates of the
ground and excited states may not be labelled by a simple
spin quantum number m_I, the spectrum becomes more complex
as more transition probabilities become non-zero. Under
these circumstances, in which the relative intensities of
the lines are not known as the eigenstates cannot be
specified, the assignment of the spectrum may be carried
out most easily by looking for repeated differences between
pairs of line positions. Each line position is subtracted

from each other line position and a table of differences
thus constructed.   This is searched until the requisite
number of pairs of differences is found; a pair of
identical differences indicates the following situation:

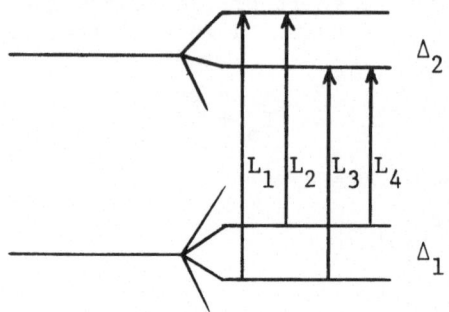

By this method, and using the fact that

$$\sum_{\substack{\text{Ground} \\ \text{state}}} E_i = \sum_{\substack{\text{Excited} \\ \text{state}}} E_j = 0$$

the energy level diagram may be constructed even when η is
not small.   It should be noted that all of the lines must
be assigned unambiguously before the analysis can proceed.
For poorly resolved spectra this could be a disadvantage,
although not all of the lines need be present or resolvable.

Quadrupole Hyperfine Parameters for ^{129}I

As an example of the use of the method, its successful
application to a Mössbauer study of IBr reported in the
literature will be described.   In a recent paper, Pasternak
and Sonninno [10] considered the Mössbauer spectrum of ^{129}I
in some interhalogen compounds.   One of these spectra, that
of IBr at 80°K is considered here.   The hyperfine parameters
η, $\mu_{\frac{7}{2}}$, $\mu_{\frac{5}{2}}$ and δ are calculated.   Pasternak and Sonninno
calculated the parameters by using an expression for the
energy levels derived by Bersohn [11] in the form of an
infinite power series in η (valid for small values of η).
A previously determined value of 1.231 was assumed for the
ratio of the quadrupole moments, Qex/Qgr.   As will be seen,
an accurate value of both η and Qex/Qgr can be determined

from one Mössbauer spectrum.

The Mössbauer transition is $\frac{7}{2} \to \frac{5}{2}$, so that for $\eta = 0$ or small, an eight line spectrum should be observed. For the reported spectrum of IBr [10], the following energy level diagram (referred to $^{e}E_a$ and $^{g}E_a$) is constructed using measured line positions.

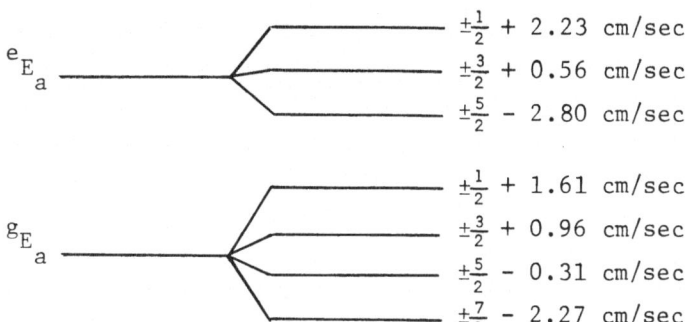

$$^{e}E_a$$

$\pm\frac{1}{2}$ + 2.23 cm/sec
$\pm\frac{3}{2}$ + 0.56 cm/sec
$\pm\frac{5}{2}$ − 2.80 cm/sec

$$^{g}E_a$$

$\pm\frac{1}{2}$ + 1.61 cm/sec
$\pm\frac{3}{2}$ + 0.96 cm/sec
$\pm\frac{5}{2}$ − 0.31 cm/sec
$\pm\frac{7}{2}$ − 2.27 cm/sec

Using equation (32) for the $(\frac{1}{2} \to \frac{1}{2})$ transition

$$0.75 = (2.23 - 1.61) + \delta$$
$$\therefore \quad \delta = 0.13 \text{ cm/sec } (0.123 \text{ cm/sec})$$

Values in parentheses after each parameter are these quoted by Pasternak and Sonninno.

For the ground $(I = \frac{7}{2})$ state, equations (18) and (19) become:

$$\mu_{\frac{7}{2}}(3 + \eta^2) = 0.0348; \quad \mu_{\frac{7}{2}}^{3}(1 - \eta^2) = -0.00126$$

Solving for $\mu_{\frac{7}{2}}$ gives: $\mu_{\frac{7}{2}} = -0.108$ cm/sec,

and $e^2 q_{\frac{7}{2}} Q_{\frac{7}{2}} = -9.07$ cm/sec $= -2090$ Mc/sec (2014 Mc/sec)

For the excited $(I = \frac{5}{2})$ state, equations (12) and (13) become:

$$\mu_{\frac{5}{2}}^{2}(3 + \eta^2) = 0.234; \quad \mu_{\frac{5}{2}}^{3}(1 - \eta^2) = -0.0219$$

This gives $e^2 q_{\frac{5}{2}} Q_{\frac{5}{2}} = -11.2$ cm/sec. Since $q_{\frac{7}{2}} = q_{\frac{5}{2}}$,

$$\frac{Q_{\frac{5}{2}}}{Q_{\frac{7}{2}}} = \frac{11.2}{9.07} = +1.23 \text{ (compared with 1.231 given previously).}$$

Finally $\eta_{\frac{5}{2}} = \eta_{\frac{7}{2}} = 0.00$ (0.06 ± 0.02).

It is felt that within the limits of accuracy imposed by the measurement of the line positions from the illustration in the literature, these results are in good agreement with those previously obtained and serve to illustrate the method.

THE ANALYTICAL DETERMINATION OF THE HYPERFINE PARAMETERS FOR A GENERAL MAGNETIC DIPOLE AND ELECTRIC QUADRUPOLE INTERACTION

In this section, the general method for determining the hyperfine parameters described previously is extended to the case of a nucleus in a general field system comprising a magnetic field vector at any orientation with respect to the principal axes of an electric field gradient tensor. The analysis follows very closely that described in the previous section with the difference that the Hamiltonian in the general case, and consequently the secular determinant and the polynomial obtained on expansion, are considerably more complex.

The simplest form of the Hamiltonian describing the combined interaction is obtained by working within the coordinate system defined by the principal axes of the E.F.G. tensor. This Hamiltonian can be written [2]:

$$H = -g_n \beta_n H[\frac{\cos\alpha}{2}(I_+ + I_-) - \frac{i\,\cos\beta}{2}(I_+ - I_-) + I_z \cos\gamma]$$

$$+ \frac{e^2 qQ}{4I(2I-1)}[3I_z^2 - I(I+1) + \frac{\eta}{2}(I_+^2 + I_-^2)]$$

$$= -A[\frac{\cos\alpha}{2}(I_+ + I_-) - \frac{i\,\cos\beta}{2}(I_+ - I_-) + I_z\cos\gamma]$$

$$+ \frac{B}{4I(2I-1)}[3I_z^2 - I(I+1) + \frac{\eta}{2}(I_+^2 + I_-^2)] \qquad (33)$$

where $A = g_n \beta_n H$ and $B = e^2 qQ$

From this Hamiltonian, the Hamiltonian matrix over the basis set $|m_i\rangle$ is constructed for nuclear spins 1, $\frac{3}{2}$, 2 and $\frac{5}{2}$. Writing the direction cosines as ℓ, m, n, the Hamiltonian matrices and secular determinants are constructed. The secular determinant for $I = \frac{5}{2}$ is given in the Appendix.

These determinants are expanded to give secular polynomials of the form:

$$E^n + a_{n-1} E^{n-1} + \ldots + a_i E^i \ldots + a_1 E + a_o = 0$$

where the coefficients a_i are complex functions of the hyperfine parameters. It should be noted that there are five independent parameters to be calculated in the most general case, i.e. A, B, η, ℓ and m (or n). The centre shift can be calculated independently after the energy level diagram is constructed.

For I = 1, the coefficients in the polynomial (equation 24) are:

$$a_2 = 0 \qquad\qquad a_1 = -A^2 - \frac{B^2}{16}(3 + \eta^2)$$

$$a_o = \frac{B}{4}\{A^2[\ell^2(3 - \eta) + m^2(3 + \eta) - 2] + \frac{B^2}{8}(1 - \eta^2)\} \qquad (34)$$

In this case, there are insufficient equations to determine the hyperfine parameters uniquely. If it can be assumed that $\eta = 0$ and that the magnetic field is parallel to the principal axes of the E.F.G. tensor, then A and B may be determined. This, however, is the trivial case in which m_I is a good quantum number and the energy levels are given by the diagonal elements of the Hamiltonian matrix.

For I = $\frac{3}{2}$, the coefficients in the polynomial

$$E^4 + a_3 E^3 + a_2 E^2 + a_1 E + a_o = 0 \text{ are:}$$

$$a_3 = 0 \qquad a_2 = -\frac{5}{2}A^2 - \frac{B^2}{24}(3 + \eta^2) \qquad\qquad (35.1)$$

$$a_1 = \frac{A^2 B}{2}[\ell^2(3 - \eta) + m^2(3 + \eta) - 2] \qquad\qquad (35.2)$$

$$a_o = \frac{9A^4}{16} + [\frac{B^2}{48}(3 + \eta^2)]^2 - \frac{A^2 B^2}{16}[4 - \frac{1}{2}(3 + \eta^2)$$

$$- \ell^2(3 - \eta)(1 + \eta) - m^2(3 + \eta)(1 - \eta)] \qquad\qquad (35.3)$$

These coefficients are then equated to the appropriate coefficients in:

$$E^4 - E^3 \sum_i E_i + E^2 \sum_{i>j} E_i E_j - E \sum_{i>j>k} E_i E_j E_k + E_1 E_2 E_3 E_4 = 0 \qquad (36)$$

Williams and Bancroft [9] have shown that from a general Fe^{57} spectrum the energy level diagram may be drawn up,

and the E_i values easily obtained. The ground state
nucleus has spin $\frac{1}{2}$, and the magnetic splitting of the
ground state is $gA = g_{g_n}\beta_n H$. Knowing the ratio of the
gyromagnetic ratios of the ground and excited states
e_{g_n}/g_{g_n}, eA, the parameter defining the magneti interaction
for the excited state is given by: $eA = (e_{g_n}/g_{g_n})gA$. There
are then three equations (36) available to obtain the four
remaining parameters β, η, ℓ and m. Consequently if one
of these is known, or can be estimated, the remaining three
can be determined analytically. The particular example
of an axial E.F.G. tensor will be mentioned here; if $\eta = 0$
equations 35.1 and 35.2 become:

$$a_2 = -\tfrac{5}{2}A^2 - \frac{B^2}{8} \text{ and } a_1 = \frac{A^2B}{2}(1 - 3n^2) \tag{37}$$

Thus A, B and γ, the angle between the magnetic field and
the axis of the axial E.F.G. tensor may be determined.

For I = 2, the coefficients in equation (29) are:

$$a_4 = 0 \qquad a_3 = -5A^2 - \frac{7B^2}{192}(3 + \eta^2) \tag{38.1}$$

$$a_2 = -\frac{B}{4}\{\frac{B^2}{64}(1 - \eta^2) + \frac{7A^2}{2}[2 - \ell^2(3 - \eta) - m^2(3 + \eta)]\} \tag{38.2}$$

$$a_1 = 4A^4 + \tfrac{3}{4}[\frac{B^2}{48}(3 + \eta^2)]^2 - \frac{A^2B^2}{8}[4 - \ell^2(3 - \eta)(1 + \eta)$$
$$- m^2(3 + \eta)(1 - \eta) - \tfrac{2}{3}(3 + \eta^2)] \tag{38.3}$$

$$a_o = B\{\tfrac{1}{3}(\tfrac{B}{8})^4(3 + \eta^2)(1 - \eta^2) - \frac{A^2B^2}{16}$$
$$+ \frac{A^2B^2}{384}(3 + \eta^2)[4 + \ell^2(3 - \eta) + m^2(3 + \eta)]$$
$$+ \frac{A^4}{2}[2 - \ell^2(3 - \eta) - m^2(3 + \eta)]\} \tag{38.4}$$

Here again there are insufficient equations available
to be able, even in principle, to determine the hyperfine
parameters (A, B, η, ℓ and m) from this single state,
without assuming a value for one of the parameters on other
grounds. As was seen with [57]Fe above, it is sometimes
possible to use the ground and excited nuclear states in
conjunction to extend the set of equations available for
the determination of the hyperfine parameters, bearing in
mind that electronic (rather than nuclear) parameters such

as η and the direction cosines ℓ, m and n are unaffected by a nuclear transition. Unfortunately, the transitions in the most important Mössbauer nuclei involving a spin state 2 (e.g. ^{182}W and ^{160}Dy) are $0 \to 2$ transitions, and consequently no information can be derived from the ground state. However, it is often possible again to assume that $\eta = 0$ as in the ^{57}Fe case considered earlier. For this case, A, B and n can be determined analytically.

For $I = \frac{5}{2}$, the coefficients a_i in the polynomial

$$E^6 + a_5 E^5 + a_4 E^4 + a_3 E^3 + a_2 E^2 + a_1 E + a_0 = 0 \tag{39}$$

are: $a_5 = 0 \quad a_4 = -\frac{7}{8}[10A^2 + \frac{B^2}{25}(3 + \eta^2)]$ \hfill (40.1)

$$a_3 = -\frac{B}{10}\{\frac{B^2}{20}(1 - \eta^2) + 14A^2[2 - \ell^2(3 - \eta) - m^2(3 + \eta)]\} \tag{40.2}$$

$$a_2 = \frac{259A^4}{16} + [\frac{7B^2}{400}(3 + \eta^2)]^2 - \frac{81A^2 B^2}{400}[4 - \frac{68}{81}(3 + \eta^2) - \ell^2(3 - \eta)(1 + \eta) - m^2(3 + \eta)(1 - \eta)] \tag{40.3}$$

$$a_1 = B\{\frac{7}{2}(\frac{B^2}{200})^2(3 + \eta^2)(1 - \eta^2) - \frac{23A^2 B^2}{200}$$
$$+ \frac{A^2 B^2}{800}(3 + \eta^2)[19 + 2\ell^2(3 - \eta) + 2m^2(3 + \eta)]$$
$$+ \frac{11A^4}{4}[2 - \ell^2(3 - \eta) - m^2(3 + \eta)]\} \tag{40.4}$$

$$a_0 = \frac{1}{20}(\frac{B^2}{20})^3(1 - \eta^2)^2 - (\frac{3AB^2}{40})^2 - \frac{225}{64}A^6$$
$$+ \frac{A^2 B^4}{640}(1 - \eta^2)[\ell^2(3 - \eta) + m^2(3 + \eta) + \frac{9}{40}(1 + \eta^2) - \frac{26}{40}]$$
$$+ \frac{A^4 B^2}{16}[(3 - \ell^2(3 - \eta) - m^2(3 + \eta))^2 - \frac{1}{2} + \frac{9}{8}(1 - \eta^2)$$
$$- \frac{\ell^2}{4}(3 - \eta)(1 + 9\eta) - \frac{m^2}{4}(3 + \eta)(1 - 9\eta)] \tag{40.5}$$

By equating these coefficients with the appropriate products of the energy levels, sufficient equations are available to be able, in principle, to determine A, B, η, ℓ and m. However, the simultaneous solution of the five equations would obviously be extremely difficult, and it is necessary to assume, or obtain via the other Mössbauer state, one or more of the parameters.

QUADRUPOLE AND MAGNETIC HYPERFINE PARAMETERS FOR ^{237}Np

In a recent study, Stone and Pillinger [4] obtained hyperfine parameters from ^{237}Np Mossbauer spectra. The transition is $\frac{5}{2} \rightarrow \frac{5}{2}$, and for $\eta = 0$ or small, sixteen lines are observed in the magnetic case.

Using line positions derived from the NpCl$_4$ spectrum at 4°K*, the coefficients in equations 40.1, 40.2 and 40.3 are used to obtain accurate analytical values of A and B for both the ground and excited states, γ (the angle between the magnetic field vector and the principal axis of the E.F.G. tensor), ρ (the ratio of the excited to ground state magnetic moments) and κ (the ratio of the quadrupole moments).

Using the method of repeated differences, the energy level diagram may be constructed as follows:

Energy levels referred to $^{e}E_a$ and $^{g}E_a$
(in cm/sec)

m_I	Excited state position	m_I	Ground state position
$-\frac{5}{2}$	+ 5.861	$-\frac{5}{2}$	+ 11.297
$-\frac{3}{2}$	+ 3.913	$-\frac{3}{2}$	+ 7.175
$-\frac{1}{2}$	+ 1.668	$-\frac{1}{2}$	+ 2.756
$+\frac{1}{2}$	- 0.874	$+\frac{1}{2}$	- 1.962
$+\frac{3}{2}$	- 3.715	$+\frac{3}{2}$	- 6.977
$+\frac{5}{2}$	- 6.854	$+\frac{5}{2}$	- 12.290

The centre shift is obtained immediately using equation (32) i.e. L_k = C.S. + ($^{e}E_j - ^{g}E_i$). For the $-\frac{5}{2} \rightarrow -\frac{3}{2}$ transition which lies at −7.166 mm/sec:

$$-7.166 = C.S. + (3.913 - 11.297)$$

and C.S. = + 0.22 cm/sec, which agrees with the value previously quoted [4]. With $\eta = 0$ for a tetragonal structure, equations 40.1, 40.2 and 40.3 become

$$a_4 = \sum_{i>j} E_i E_j = -\frac{7}{8}[10A^2 + \frac{3B^2}{25}] \tag{41.1}$$

$$a_3 = -\sum_{i>j>k} E_i E_j E_k = -\frac{B}{10}[\frac{B^2}{20} + 14A^2(3n^2 - 1)] \tag{41.2}$$

$$a_2 = \sum_{i>j>k>\ell} E_i E_j E_k E_\ell = (\frac{21B^2}{400})^2 + \frac{259A^4}{16} - \frac{81A^2B^2}{400}[3n^2 - \frac{41}{27}] \tag{41.3}$$

Calculating the products of the roots from the energy level diagram gives the following results:

$^e a_4 = -56.99$, $^e a_3 = +35.98$, $^e a_2 = +699.05$, $^g a_4 = -195.13$, $^g a_3 = +123.74$ and $^g a_2 = +7989.97$. Solving the above equations, and taking the realistic roots gives the following values for the hyperfine parameters. The values in parentheses are those given previously [4].
$g_B = -1.98(-1.68)$, $e_B = -1.99(-1.68)$, $g_A = +4.72(+4.74)$ $e_A = +2.54(+2.54)$, $\cos\gamma = 1.00$ (assumed), $K = +1.01(+1.00)$ and $\rho = +0.54(+0.537)$. With the exception of the values of B, these parameters are in excellent agreement with those quoted previously. That $\gamma = 0°$ is confirmed and an accurate value for Q^{ex}/Q^{gr} is obtained.

Having confirmed that $\cos\gamma = 1$, it is apparent that the hyperfine parameters may be obtained very simply from the diagonal elements of the Hamiltonian matrix and the energy level diagram given previously.

CONCLUSION

If sufficient line positions can be derived from an observed Mossbauer spectrum, the above analytical treatment is very useful for deriving the hyperfine parameters in a comparatively simple fashion.

However, the estimation of errors in the derived parameters cannot be carried out easily. In the previous NpCℓ_4 example, the errors - at least for A and B - are of the same order of magnitude as the errors in the line positions. In some cases, it may be preferred to use the derived parameters as initial estimates in a least squares process.

APPENDIX

Secular determinant for the quadrupole hyperfine interaction for $I = \frac{5}{2}$, where $\mu = \dfrac{e^2 qQ}{4I(2I-1)}$

$$
\begin{vmatrix}
10\mu_{\frac{5}{2}}-E & 0 & \sqrt{10}\mu_{\frac{5}{2}}\eta & 0 & 0 & 0 \\
0 & -2\mu_{\frac{5}{2}}-E & 0 & \sqrt{18}\mu_{\frac{5}{2}}\eta & 0 & 0 \\
\sqrt{10}\mu_{\frac{5}{2}}\eta & 0 & -8\mu_{\frac{5}{2}}-E & 0 & \sqrt{18}\mu_{\frac{5}{2}}\eta & 0 \\
0 & \sqrt{18}\mu_{\frac{5}{2}}\eta & 0 & -8\mu_{\frac{5}{2}}-E & 0 & \sqrt{10}\mu_{\frac{5}{2}}\eta \\
0 & 0 & \sqrt{18}\mu_{\frac{5}{2}}\eta & 0 & -2\mu_{\frac{5}{2}}-E & 0 \\
0 & 0 & 0 & \sqrt{10}\mu_{\frac{5}{2}}\eta & 0 & 10\mu_{\frac{5}{2}}-E
\end{vmatrix} = 0
$$

Secular Determinant for the Combined Magnetic–quadrupole Interaction for $I = \frac{5}{2}$

$-\frac{5nA}{2}+\frac{B}{4}-E$	$-\frac{A\sqrt5(1-im)}{2}$	$\frac{nB\sqrt{10}}{40}$	0	0	0	
$-\frac{A\sqrt5(1+im)}{2}$	$-\frac{3nA}{2}-\frac{B}{20}-E$	$-A\sqrt2(1-im)$	$\frac{3nB\sqrt2}{40}$	0	0	
$\frac{nB\sqrt{10}}{40}$	$-A\sqrt2(1+im)$	$-\frac{nA}{2}-\frac{B}{5}-E$	$-\frac{3A(1-im)}{2}$	$\frac{3nB\sqrt2}{40}$	0	$= 0$
0	$\frac{3nB\sqrt2}{40}$	$-\frac{3A(1+im)}{2}$	$\frac{nA}{2}-\frac{B}{5}-E$	$-A\sqrt2(1-im)$	$\frac{nB\sqrt{10}}{40}$	
0	0	$\frac{3nB\sqrt2}{40}$	$-A\sqrt2(1+im)$	$\frac{3nA}{2}-\frac{B}{20}-E$	$-\frac{A\sqrt5(1-im)}{2}$	
0	0	0	$\frac{nB\sqrt{10}}{40}$	$-\frac{A\sqrt5(1+im)}{2}$	$\frac{5nA}{2}+\frac{B}{4}-E$	

REFERENCES

1. J.R. Gabriel and S.L. Ruby, Nucl. Inst. Methods 36 23 (1965).

2. G.R. Hoy and S. Chandra, J. Chem. Phys. 47, 961 (1967).

3. W. Kundig, Nucl. Instr. Methods 48, 219 (1967).

4. A.J. Stone and W.L. Pillinger, Phys. Rev. #4 165, 1319 (1968).

5. P.G.L. Williams and G.M. Bancroft, Mol. Phys. 19, 717 (1970).

6. C.P. Slichter, The Principles of Magnetic Resonance, Harper and Row (1964).

7. M.H. Cohen, Phys. Rev. 96, 1278 (1954).

8. Handbook of Mathematical Functions, edited by M.A. Abramowitz and I.A. Stegun, National Bureau of Standards, Applied Mathematics Series 55, 17 (1964).

9. P.G.L. Williams and G.M. Bancroft, Chem. Phys. Lett., 3, 110 (1969).

10. M. Pasternak and T.J. Sonnino, J. Chem. Phys. 48, 1997 (1968).

11. R. Bersohn, J. Chem. Phys. 20, 1505 (1952).

12. C.E. Bemis and K. Fransson, Phys. Lett. 19, 567 (1965).

THE MÖSSBAUER EFFECT OF Fe^{57} IN RARE GAS MATRICES AT 4.2K*

T. K. McNab† and P. H. Barrett

Department of Physics
University of California
Santa Barbara, California 93106

1. INTRODUCTION

Since the late nineteen fifties optical spectroscopists have developed the technique of rare gas matrix isolation [1] in order to obtain data on individually trapped atoms, molecules and radicals at temperatures of 20K or less. The technique presented several advantages over observing the same materials in the gas phase; the most important being that normally short lived molecules and molecular fragments could be frozen into a rare gas matrix for long periods of time and detailed spectra taken. Experiments on atoms in both the gaseous and trapped phases indicated that the solvent rare gas did not affect appreciably the optical properties of the atoms, thereby establishing the veracity of the method.

Recently, several investigators have used the Mössbauer effect technique to study the hyperfine interactions of Kr^{83} in solid krypton [2]; however, these are not rare gas matrix isolation experiments since an impurity

* This work was supported by the U.S. Atomic Energy Commission.

† Now at Argonne National Laboratories, Argonne, Illinois 60440

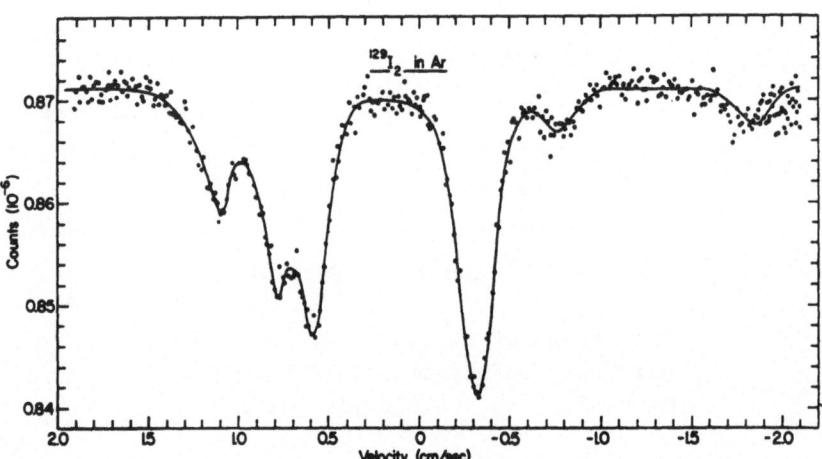

Fig. 1. Spectrum of I_2 in solid argon (after Bukshpan, et al. [3]). It was prepared by mixing argon gas with iodine vapor at room temperature and freezing the mixture at 22K.

atom is not being studied. The first successful rare gas matrix isolation experiment was carried out with iodine molecules (I_2) imbedded in a solid argon matrix at T=22K [3]. The results obtained in this experiment indicated that the isolated molecule, indeed can be considered as an almost free entity where matrix interactions are negligible. Further, the mean square displacement of the iodine molecule is sufficiently small to give rise to an appreciable recoil free fraction. Here the effect was approximately 4% using a $Zn^{129}Te$ source at 100K, as shown in Fig. 1. The absorber consists of 20 mg/cm^2 of I_2 in argon with a molecular ratio Ar/I_2=100.

It is the intent of this paper to describe a matrix isolation apparatus, and the necessary experimental procedures, for Mössbauer nuclei which must be evaporated from a furnace; then to report the results that have been obtained with atomic concentrations ($\lesssim$ 3%) of iron metal dissolved in solid Ar, Kr and Xe matrices at T=4.2K.

2. GENERAL CONSIDERATIONS

In an experiment where the Mössbauer nuclei normally form a solid with a low vapor pressure ($\lesssim 10^{-4}$ torr) at 300K,

elevated temperatures are required in order to create the
necessary flux of atoms or molecules for injection into the
rare gas matrix. Such is the case with most Mössbauer
nuclei and consequently an oven or furnace system is required.

For many years optical spectroscopists have used
Knudsen cells for this purpose since they can produce temperatures sufficient to evaporate most metals. The advantage to be gained in using the cell is that the atoms are
in equilibrium and it is possible to calculate the atomic
flux knowing the temperature and the channel dimensions.
Mössbauer effect experiments, however, require 10^2-10^3 times
the number of atoms necessary for optical experiments; thus
the orifice dimensions would have to be so large that
equilibrium would never be established and the primary
advantage of the cell lost.

It is advantageous to utilize instead, an open-ended
cylindrical crucible for atoms such as iron and tin. For
example, an alumina crucible, 18 mm long and 3 mm I.D.
placed 8 or 9 cm from a cold substrate will produce an
absorber 15 mm in diameter with an evaporation efficiency
(absorber weight per cm^2/total weight evaporated) $\approx 2\%/cm^2$.
This figure is an order of magnitude improvement over
simple 2π evaporation.

A preliminary description of such a system has been
published [4] already. A diagrammatic plan view is shown in
Fig. 2. It is drawn approximately to scale and an idea of
the dimensions can be obtained from the 1 inch window in the
proportional counter. The crucible is inserted into a 5 mm
I.D. tantalum furnace F which can be resistance heated to
$\approx 1400^\circ$C. A close fitting tantalum shield reflects radiant
energy back to the furnace, and an outer water-cooled copper
shield removes excess heat up to a rate ≈ 350 watts.

The furnace housing H is attached to the side of a
liquid helium cryostat by means of a demountable "O" ring
seal. Thus material evaporated from the crucible will con-
dense onto a 1 1/8 inch beryllium disk B, which is held in

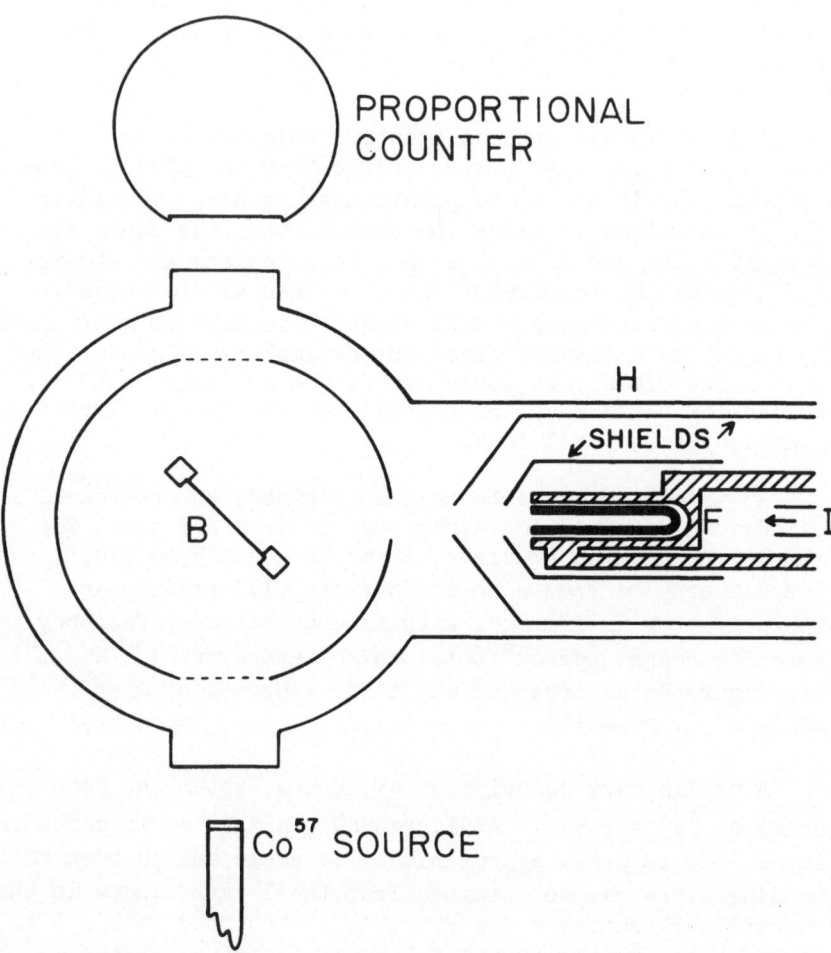

Fig. 2. A diagrammatic plan view of the furnace and substrate used for matrix isolation experiments. The geometry for matrix thickness determination and Mössbauer data accumulation is shown.

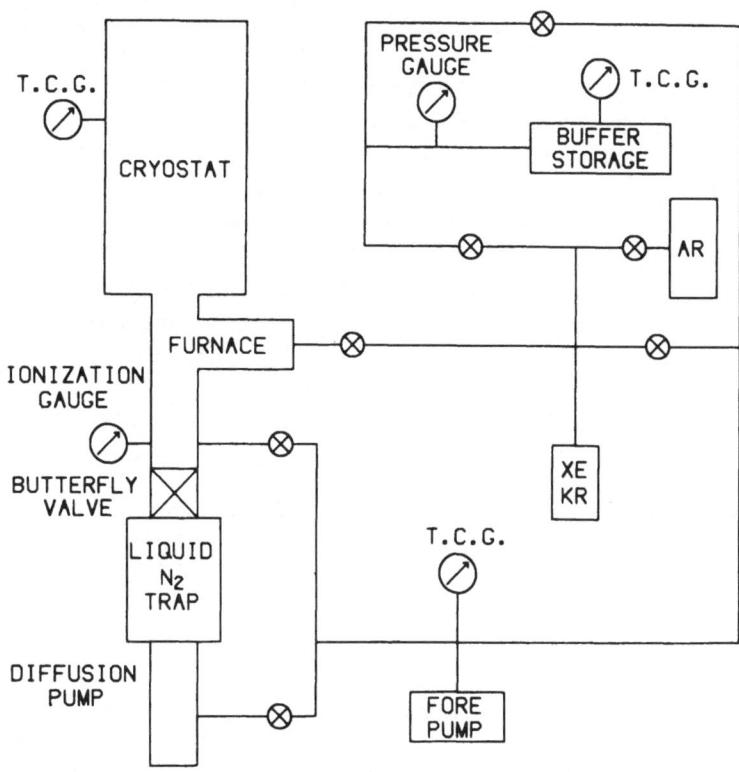

Fig. 3. Block diagram of the cryostat and vacuum and gas handling system used in the present work.

an oxygen free copper clamp thermally bonded to the bottom of the liquid He⁴ bath.

In order to isolate iron atoms (we shall refer only to iron hereafter although the discussion will be applicable to many other atomic species also) a stream of rare gas with an adjustable flow rate enters through I, and condenses onto the cooled Be disk.

Figure 3 is a diagrammatic representation of the cryostat, vacuum and gas handling system used in the present work. The stainless steel cryostat sits on a rotatable "O" ring seal so that outgassing of the furnace can be carried out before exposing the beryllium substrate to the rare

gas/Fe mixture. The diffusion pumping system has a speed
of 35 ℓ per second which is sufficient to reduce the
cryostat pressure from 10^{-1} torr to $\approx 10^{-6}$ torr in 12 hours
without the use of the cold trap. When liquid N_2 is added
to the cold trap and cryostat the pressure will then drop
below 2×10^{-7} torr in 30 minutes. Finally with liquid
helium in the cryostat the pressure indicated on the ionization gauge $\approx 8 \times 10^{-8}$ torr. This low pressure is absol-
utely necessary since even under these conditions contam-
inant gas will accrete on the absorber at a rate which is
10%-20% that for iron atoms in a typical run.

The gas handling lines are constructed of stainless
steel tubing and are pumped continuously for 12 hours before
each deposition run. The rare gases used in our experiments
have only a few parts per million impurities.

3. OPERATING PROCEDURES

After the furnace has been outgassed and allowed to
cool, the cryostat is rotated under vacuum so that the
three beam apertures are concentric. Initially, a layer of
rare gas (50 μg thick) is laid down on the Be in order to
eliminate iron-substrate interactions. The furnace is
raised to the required temperature which, in conjunction
with the flow rate, controls the rare gas-to-iron atomic
ratio. The temperature is held to within $\pm 5^{\circ}C$ and is read
with an optical pyrometer via a window positioned at the
rear of the furnace. The dimensions of the crucible dictate
the rate at which iron can be deposited. It is required
that the iron gas atoms have a mean free path greater than
the crucible length otherwise they will be scattered out of
the beam. This means that the iron vapor pressure $\lesssim 10^{-2}$
torr for efficient deposition. For an absorber consisting
of 1-2% Fe, the rare gas enters the cryostat at a rate
$\approx$ 1cc/STP per minute; a total sticking efficiency for the
rare gas $\approx$ 50% seems to be indicated.

When the absorber is being laid down the radiative
thermal loading on the absorber holder and beryllium sub-

strate is sufficient to boil off an extra 500 cc liquid He^4
per hour, which represents a heat input of 350 mW. This
does not affect appreciably the rate of accretion of rare
gas (except neon) until $\approx$ 5 mg/cm^2 has been deposited. For
greater matrix thicknesses the rate drops gradually;
therefore, in our experiments all absorbers consist of no
more than 5 mg/cm^2 of rare gas. After completion of the
iron deposition another layer of rare gas (50 µg thick)
covers the matrix to eliminate the possibility of impurity
gases in the vacuum interacting with near surface Fe atoms
in the matrix.

An indication of the temperature of the topmost part
(facing the furnace) of the matrix during the deposition
can be deduced from work with neon. In this latter case,
the neon will not stick at a temperature above $\approx$ 8K since
its vapor pressure becomes comparable with the deposition
pressure. It was found that this temperature was reached
for a furnace temperature $\approx$ 1200°C, which is 50 - 100°C
below those required for the Fe in rare gases experiment.
This indicates that the matrix temperatures is $\approx$ 12K, when
the furnace is run at the required temperature of 1300°C; it
is impossible therefore, to perform the implantation of Fe
in neon with our apparatus. It should be possible with Sn
however, since this metal has a vapor pressure of 10^{-2} torr
at $\approx$ 1200°C.

Approximately 3-4 hours are required to build up an
absorber and the rare-gas accretion on the beryllium disk
is continuously monitored by measuring the attenuation of
the iron 6.4 keV K X-ray from a Co^{57} in Cu source as indicated in Fig. 2. In this way, the rare gas matrix thickness
can be measured to within ±7%. If a thick matrix is to be
made then it might be advisable to use the 14.4 keV γ-ray
from Fe^{57m} or the 23.8 keV γ-ray from Sn^{119m} to determine
the matrix thickness. The half-thickness values [5] of Ne, Ar,
Kr and Xe for electromagnetic radiation between the energies
5-70 keV are given in Fig. 4.

It is not possible, however, to determine simultaneously
the deposition rate of the iron. This was carried out in a

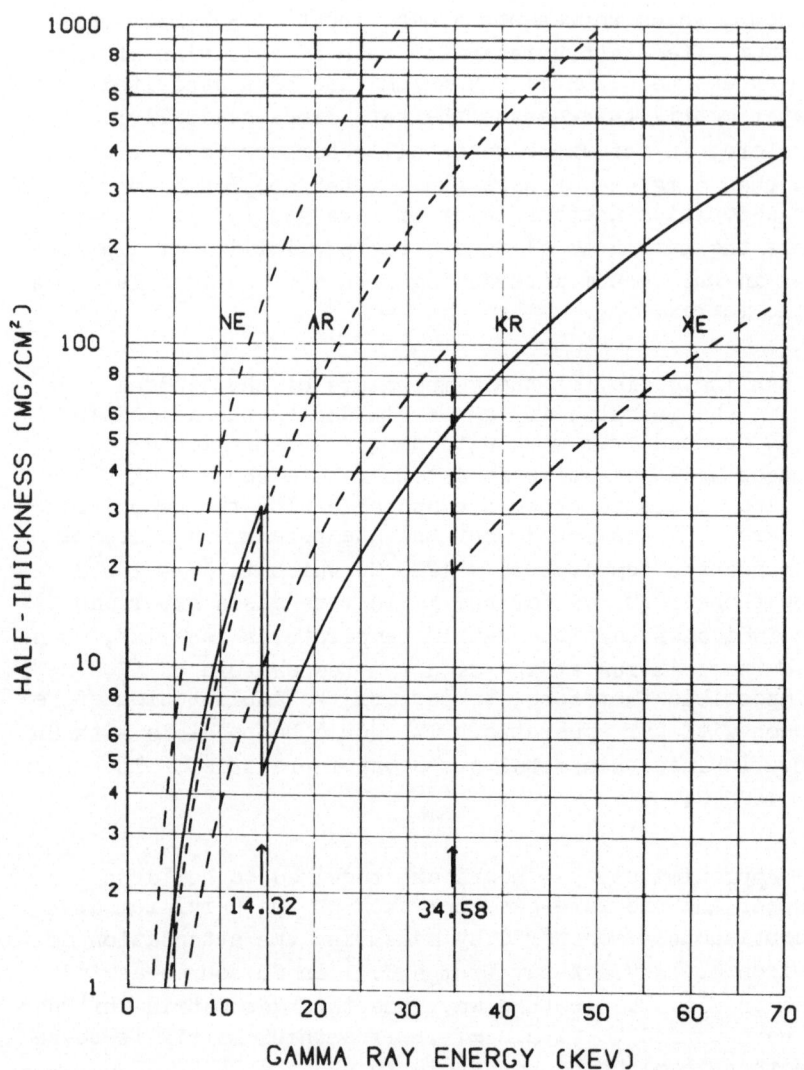

Fig. 4. Half-thickness coefficients for the four rare gases.
They are calculated from the empirical formulae given by
Leroux [5]. The values are calculated for photoelectric
absorption only, and agree with published data to within
±5%. In the region of the K-edge they may be more inaccu-
rate than this.

separate experiment where a known weight of iron was evaporated and the collection efficiency of the Be disk was
determined by weighing before and after the deposition. It
was possible using a microgram balance to weigh the deposited iron ($\approx$ 180 μg) to within a few per cent. Thus weighing
the crucible before and after a run, together with the so
determined collection efficiency, gives the weight of iron
in the absorber.

Initially, we attempted to determine the iron on the
beryllium with a Mössbauer effect measurement, but the major
drawback to this method is the lack of data on the f- factor
of iron deposited under these conditions. There is still
one major problem associated with obtaining a reliable
figure for the iron concentration /cm^2. This is the deposition of a non-uniform layer because of geometrical considerations. Attempts were made to determine the variation
in iron layer thickness across the foil by carrying out
Mössbauer effect measurements on different 3 mm diameter
areas of the disk. These data indicated that the iron concentration across the absorber is constant to within ±10%.
Taking all sources of error into account we calculate an
absolute error ±15% for the concentration of iron in rare
gas.

One seemingly trivial yet nonetheless important point
which requires attention is that the type of furnace described above is a non-equilibrium system; at any given temperature, the output from the crucible depends critically on
the exposed surface area of the iron powder. Consequently
in the time of a three hour run, the beam flux can drop as
the total quantity of metal in the crucible decreases, and
also in the case of metals below their melting point, sinters
with a concomitant reduction in surface area. The former
difficulty can be obviated almost completely by placing a
much larger charge of metal in the crucible than is required
for that particular run; the latter becomes significant after
8 or 10 hours evaporation, therefore so long as the partially
sintered mass is broken up after each run, reproducible
results will be obtained.

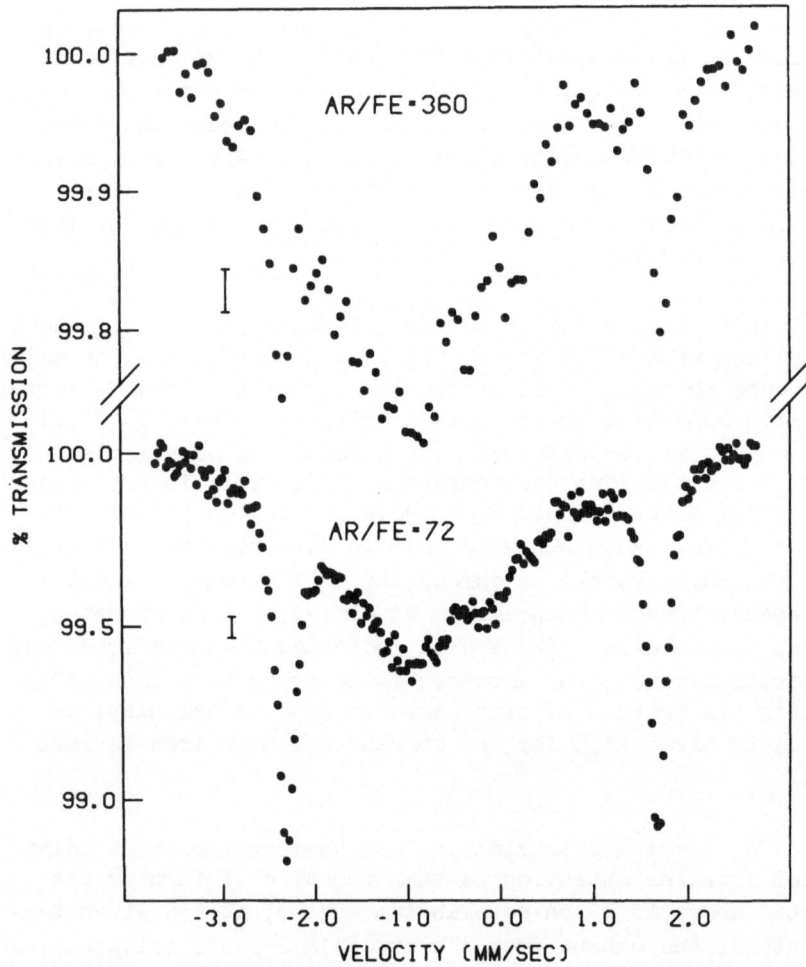

Fig. 5. Spectra of Fe^{57} in solid argon at 4.2K.

4. EXPERIMENTAL RESULTS

The Mössbauer spectra were obtained with a conventional constant acceleration spectrometer using a Co^{57} in Pd source and an enriched Fe^{57} foil for calibration purposes. The spectra exhibited in Figs. 5-7 summarize essentially all the data of interest obtained at T=4.2K, when Fe^{57} is implanted into the rare gas matrices argon, krypton and

Table I. Data for Fe^{57} in Frozen Rare Gases at T = 4.2K

Matrix	Atomic Concentration of Fe in Rare Gases	Monomer		Dimer	
		δ	Γ	Q.S.	δ
Argon	360 ± 54	-0.72±.05	2.1±.2	4.06±.05	-0.15±.02
	72 ± 11	-0.72±.03	1.6±.1	4.04±.05	-0.13±.02
Krypton	136 ± 20	-0.75±.02	0.76±.03	4.1 ±.1	-0.14±.04
	44 ± 7	-0.75±.03	0.67±.02	4.11±.08	-0.13±.03
Xenon	94 ± 14	-0.73±.02	0.82±.03	—	—
	31 ± 5	-0.72±.03	0.72±.03	—	—

The units of δ, Γ and Q.S. are mm/sec.

Isomer shift values are given with respect to an Fe^{57} metal absorber at 300K.

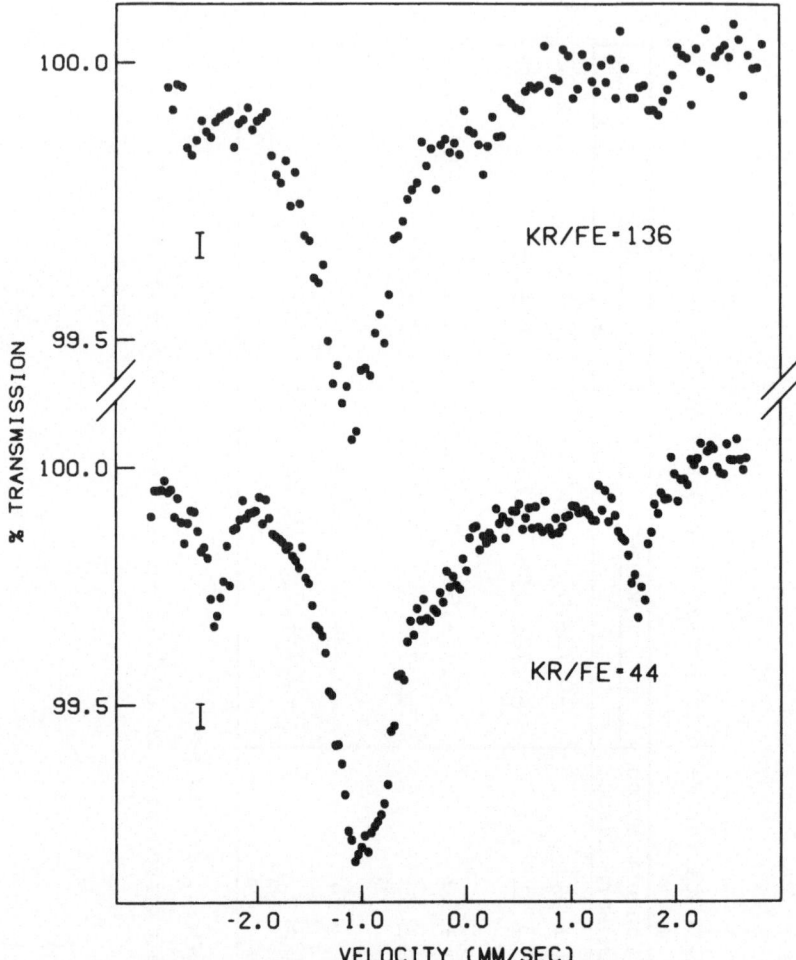

Fig. 6. Spectra of Fe^{57} in solid krypton at 4.2K.

xenon at concentrations $\leq 3\%$. In all cases, the iron in the
matrices consists of $90\% \sim Fe^{57}$, and the data is uncorrected
for non-resonant background. The atomic ratio of rare gas/
iron is given for each spectrum.

        Consider first the argon data in Fig. 5. The two
spectra represent concentrations of Fe in Ar of approximately
0.3% and 1.4%. At the lower concentration a single broad

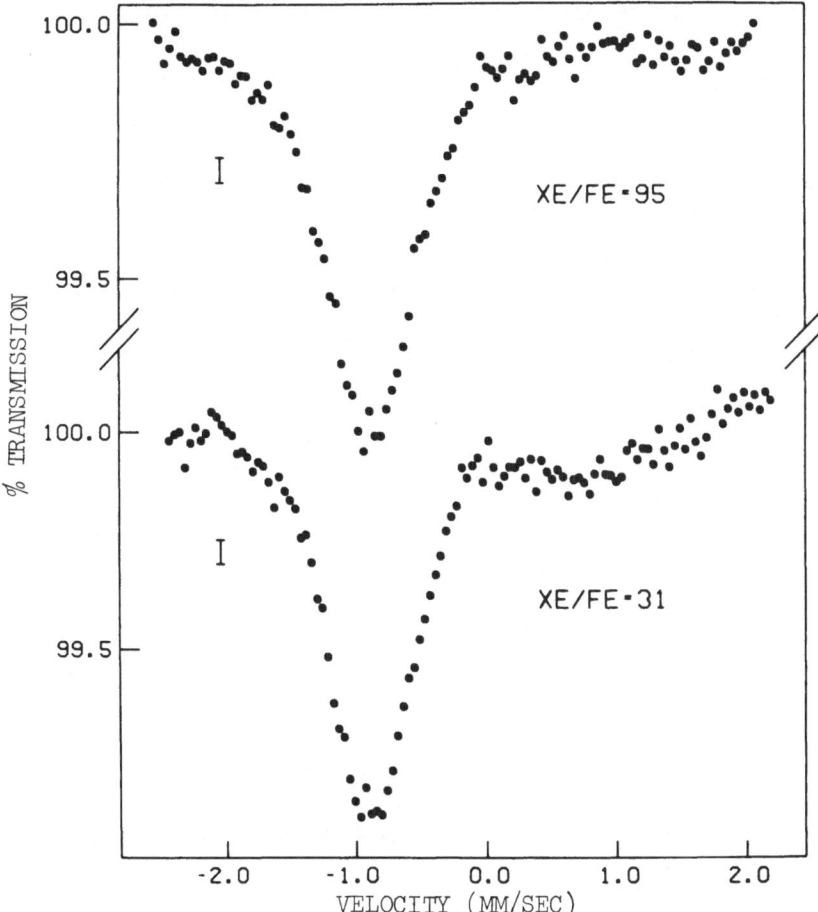

Fig. 7. Spectra of Fe^{57} in solid xenon at 4.2K.

line, with possibly some impurity contributions to the high
velocity side of the peak, dominates the spectrum; two
narrow, lower intensity lines complete the picture. When
the concentration of iron in argon is increased fivefold the
two narrow lines dominate, although the broad central line
is still extant, but now another broad feature manifests itself at ≈ 0.0 mm/sec.

On changing the matrix to krypton (see Fig. 6) there is
a marked change in the relative amplitudes of the two main
components observed in argon. At 0.74% concentration a

broad line dominates, but now the narrow peaks have disappeared almost; even increasing the amount of iron to 2.3%
does not cause the narrow lines to form a large part of the
total resonance effect. At this higher concentration the
intense line is not symmetric. It has been distorted by a
peak contribution at ≈ 0.0 mm/sec, as was the case for the
argon spectrum with Ar/Fe = 72.

Finally, the xenon spectra in Fig. 7 do not show any
traces of the narrow lines. For ≈ 1% iron in xenon a single
broad line is observed; on increasing this value to ≈ 3%,
the only change is the observation of a very broad shallow
line located at ≈ 1.0 mm/sec.

All spectra have a common feature; a broad line which
completely dominates at low concentrations in all matrices.
We ascribe this peak to the presence of isolated atoms of
Fe^{57} which do not have as any of their nearest neighbors
another iron atom. This, therefore, is the monomer with
the ground state electron configuration $3d^6 4s^2$. The narrow
line components, which intensify at higher concentrations
in Ar and Kr, thus can be interpreted as the dimer; here,
one Fe^{57} atom has as one of its nearest neighbors another
iron atom. The data for all six spectra are summarized in
Table 1. It does not include parameters on the afore-
mentioned contributions to the spectra which do not fall
into either the monomer or dimer category; these will be
discussed in section 9.

5. THE MONOMER ISOMER SHIFT

The most gratifying results are those for the isomer
shift of the monomer which is identical within the error
for each of the three matrices. This agrees well with
optical spectroscopic data as discussed initially, where
the matrix has only a minimal effect on the trapped species.
The averaged value for the isomer shift with respect to iron
metal at 300K is (-0.74 ± .02) mm/sec. The error perhaps
should be increased to ± .05 mm/sec to take into account
possible systematic errors in the isomer shift calibration.

Wertheim [6] has determined accurately the isomer shift
of FeF_2 at T = 4.2K to be (+ 1.47 ± .01) mm/sec with respect
to iron metal at 300K. Since it is the most ionic of the
ferrous compounds we shall assume (as have most other
investigators [7,8]) that its isomer shift accurately
reflects the electron configuration $3d^6 4s^0$. Therefore, on
subtracting 2 x 4s electrons from the monomer the isomer
shift is changed by +(2.21 ± .06) mm/sec. For background
information on the following calculations we refer the reader
to Danon's [8] review paper.

The difference $\Delta\delta$ between the two isomer shift values
can be equated in the following manner:

$$\Delta\delta = F(Z) \cdot \Delta\Psi^2(0) \cdot \frac{\Delta R}{R} ,$$

where $\Delta\Psi^2(0) = \Psi_A^2(0) - \Psi_B^2(0)$, and $\Psi_A^2(0)$ and $\Psi_B^2(0)$ are
respectively, the total relativistic s-electron density at
the nucleus for the ferrous iron and the monomer. $\Delta R/R$ is
the change in radius between the first excited state and the
ground state of the Fe^{57} nucleus. $F(Z)$ is a constant which
depends only on nuclear parameters and includes a correction
factor for relativistic effects.

The electron contribution to the isomer shift has been
calculated by Watson [8] using restricted Hartree-Fock wave
functions. He finds that

$$\Psi_A^2(0) = 11879.8 \text{ a.u.}$$

and

$$\Psi_B^2(0) = 11885.8 \text{ a.u.}$$

Therefore $\Delta\Psi^2(0) = 6.0$ a.u. With these values for $\Delta\delta$,
$\Delta\Psi^2(0)$ and $F(Z)$ [9] we obtain

$$\Delta R/R = -(10.2 \pm 0.3) \cdot 10^{-4} .$$

This compares with previous determinations in which
$-12.10^{-4} \le \Delta R/R \le -4.10^{-4}$ where the relativistic correction
has been applied to all earlier data.

6. THE MONOMER LINEWIDTH

Table 1 indicates that the linewidth of the monomer peak is different for each of the three rare gas matrices and varies, also to a lesser degree, with concentration for any given matrix. It is a relatively simple matter to account for this latter phenomenon (as outlined in the following) in terms of the surface trapping mechanism of the deposited atoms, together with diffusion in the case of argon.

In order to obtain higher concentrations of iron in a rare gas matrix the furnace temperature is increased and the gas flow reduced. For these conditions, the available time to trap iron and rare gas atoms onto a cubic site at the matrix surface is maximized. Lower concentration runs demand a greater flow of rare gas and the surface diffusion time is thereby reduced; this leads to the creation of a more dislocated matrix.

It is to be expected that in an imperfect rare gas lattice the non-cubic crystalline fields that are manifested at the iron atom sites will broaden the line due to their interaction with the Fe^{57m} quadrupole moment. A more dislocated environment will lead therefore to a broader line. Micklitz [10] also has observed in the optical absorption spectra of iron atoms in rare gas matrices, that narrower lines are obtained when the gas flow rate is reduced. This is the same situation as for the Mössbauer studies, since ill-defined crystalline environments give rise to additional optical line broadening.

A mechanism that allows matrix annealing manifests itself particularly, in the argon experiments. At high concentrations the radiative loading on the matrix due to the furnace may be sufficient to raise its temperature at the front surface to the point where diffusion will take place. This need be only about 14K and would promote annealing of the crystal structure. Therefore, in this case, higher concentrations will give rise to a narrower line. With the above reasoning it is possible to account for the discrepancies in linewidth between high and low concentrations of iron in a given rare gas.

We do not have sufficient data at this juncture, however, to determine the nature of the residual line broadening in each of the three matrices. It is not clear at this stage that the "annealed" linewidths for argon, krypton and xenon are different.

Mann and Broida [11] have observed the optical absorption spectra of iron atoms isolated in Ar, Kr and Xe matrices at T = 4.2K, for concentrations ≈ 0.03%. They find that the spectra for iron in Kr and Xe are similar and not very different from those in the gas phase. In argon, however, the situation was found to be more complicated with approximately double the expected number of lines appearing. This might suggest that the iron atoms occupy two different crystallographic sites in the argon matrix; this could account for the extra broadening of the line (above the Kr and Xe values) in the Mössbauer spectra if one of the sites had non-cubic symmetry.

In order to determine the origin of the line broadening we are initiating a series of experiments to ascertain the manner in which temperature influences the line width. These results will be published at a later date.

7. THE DIMER QUADRUPOLE SPLITTING

The data obtained with the three different rare gas matrices indicates that the maximum distance for interaction of the $3d^6$ electron wave functions for two iron atoms is less than the nearest neighbor xenon spacing (4·33Å) [12] but larger than the krypton spacing (3·99Å) and of course that for argon (3·75Å). A detailed computer calculation of the $3d^6$ electron shell overlap energy for two iron atoms at varying separations will be attempted in order to determine the degree of interaction necessary for the observation of the large quadrupole splitting; these results, together with a discussion of the dimer isomer shift, will be published at a later date.

The values for the quadrupole splitting and the isomer

shift of the dimer are the same within the error (see
Table 1) for both the matrices in which it exists. It is
evident therefore that when two iron atoms become sufficiently close they assume an equilibrium separation which
is matrix independent. The axial symmetry, which the
dimer exhibits, dictates the orbital ground state of the
$3d^6$ electrons and in turn the electric field gradient (EFG)
at the Fe^{57} nucleus.

Ingalls [13] has considered the effects of different
symmetries on the fivefold orbitally degenerate ^5D ground
state of the ferrous ion. He shows that under axial
symmetry the ground state degeneracy is lifted partially.
Each of these states gives rise to a free ion EFG of
$\pm(4/7)<r^{-3}>_{eff}$; the term includes antishielding corrections.
Now for Fe^{57m} with spin $3/2$, the quadrupole splitting is

$$\Delta E = \tfrac{1}{2}e^2qQ \quad ,$$

where q is the EFG and Q is the quadrupole moment of the
first excited state of Fe^{57}. Substituting for q gives

$$\Delta E = \pm(2/7)e^2<r^{-3}>_{eff} Q \quad .$$

With $\Delta E = (4.05\pm.05)$ mm/sec and the calculated value for
$<r^{-3}>_{eff} = 3.3$ a.u. [14] we obtain $Q = \pm0.213b$.

The sign of Q has been determined [15] to be positive,
therefore

$$Q = 0.213b \quad .$$

The error in this value is determined by the accuracy with
which Freeman and Watson [14] calculated $<r^{-3}>_{eff}$ for the
free ion. The experimental contribution to the error is
only about 1%.

8. THE MONOMER/DIMER RATIO

It is apparent on inspection of the argon and krypton

spectra that the ratio of monomer to dimer is matrix
dependent. In the high concentration runs the quantity of
iron in each matrix is approximately the same; for 0.74%
iron in krypton the dimer peak has disappeared almost,
whereas it is readily observed in argon even for 0.28% concentration. These results would seem to suggest one of the
two following possibilities:

(a) The f factor for the dimer in Kr is considerably
smaller than for Ar.

(b) A migration effect in argon enhances the dimer
concentration.

These two alternatives will be considered in terms of
the occupation probability of the iron atoms on rare gas
matrix crystal sites. It will be assumed that the iron
atoms are distributed in a random fashion throughout the
matrix, which for the rare gases is a fcc structure [12].
Here each lattice site has twelve nearest neighbors at a
separation $= a/\sqrt{2}$, where a is the lattice constant.

The probability P_N [16], that one iron atom has N of
its twelve nearest neighbors other iron atoms can be
expressed as

$$P_N = \frac{12! \, \varphi^N (1-\varphi)^{12-N}}{(12-N)! \, N!} \quad ,$$

where φ is the fractional atomic concentration of solute.
This gives the probability for the monomer, $P_0 = (1 - \varphi)^{12}$,
and for the dimer $P_1 = 12\varphi(1-\varphi)$. These functions are
graphed in Fig. 8 for the concentration regime of interest;
also included is P_2 since it will be referred to later in
section 9.

With these formulae the probability of occurrence of
monomer and dimer are computed for the four argon and
krypton spectra. These values are given in Table II; also
given is the calculated ratio of the dimer to monomer and
the experimentally determined ratio. In krypton the calculated and experimental values are approximately equal
whereas in argon there appear to be three times the number

Table II. Calculated and Experimental Dimer/Monomer Ratios (D/M)

Matrix	φ	P_o	P_1	P_2	Dimer/Monomer Calculated	Dimer/Monomer Expt^l	$\left\{\dfrac{(D/M)Calc}{(D/M)Expt}\right\}$
Argon	.014	.85	.14	.01	.165	.50	.33 ± .03
	.0028	.965	.035	.00	.036	.10	.36 ± .05
Krypton	.023	.76	.21	.03	.276	.27	1.0 ± .2
	.0074	.915	.085	.00	.093	.085	1.1 ± .3

ATOMIC RATIO OF RARE GAS TO IRON

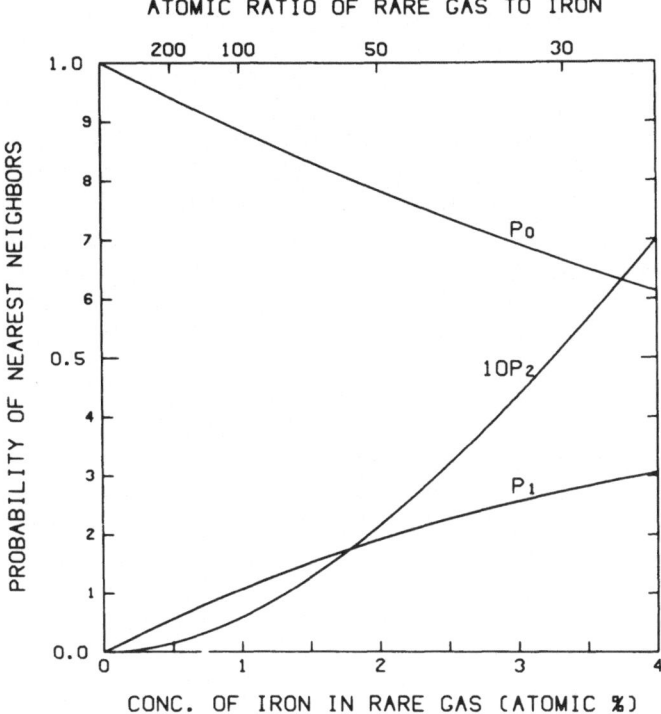

Fig. 8. Probability of occurrence of monomer, dimer and trimer for iron atoms distributed at random on solid rare gas crystal sites.

of dimers expected.

On examination of these results in terms of (a) or (b) it would appear that migration of iron in argon is more likely than a factor of 3 drop in f-factor for the dimer from argon to krypton. The manner in which the argon matrix is fabricated is further support for migration. It has been pointed out already (see section 6) that the argon becomes sufficiently warm for annealing to take place. In this temperature regime it would be hardly surprising to find that iron migration can take place.

9. IMPURITY LINES AND DISTORTION EFFECTS

Impurity lines here are defined as all those spectral contributions which do not fall into either the monomer or dimer category. It must be mentioned however, that one contribution to the spectra in Figs. 5-7 has been removed. This is the impurity spectrum derived from the two beryllium disks one of which is the substrate and the other is the proportional counter window. Together, they give rise to an effect $\approx$ 0.06% which is appreciable for spectra like the low concentration iron in argon.

The "impurities" fall into several categories. Consider first the dilute Fe/Ar (Fig. 5); it is possible that there are as many impurity atoms and molecules (O_2, N_2 and furnace outgassing products) in the matrix as iron atoms. The probability of an iron atom having one of these molecules as a nearest neighbor is a few per cent. Under these circumstances it is to be expected that a small impurity will appear in the spectrum. Identification may be possible after the Mössbauer spectra of FeO_2, FeN_2 have been observed by themselves in an argon matrix.

In the high concentration Fe/Ar matrix the probability P_2 (see Fig. 8) for trimers is 1.0%; it seems likely however, that this number should be 3 or 4 times larger since diffusion appears to take place fairly readily in argon when the matrix is being made. The impurity at 0.0 mm/sec therefore, is identified tentatively with the presence of the trimer. We are attempting to increase the amplitude of this line at present, so that its characteristics can be determined more readily.

The trimer peak appears in the concentrated krypton spectrum also; its effect being to distort the high velocity side of the monomer peak. On reduction of the iron content in the matrix this peak disappears.

A different type of impurity is seen in the xenon spectrum for Xe/Fe=31. This run represents an attempt at making the highest possible concentration of iron in a rare

gas without the formation of iron clumps. It was partially
successful only. The reason is that it is difficult to
obtain a uniform gas flow at small input rates and therefore the iron tends to clump at times. In this condition,
the iron gives rise to a magnetic hyperfine spectrum which
is characteristic of bulk iron metal but with extremely
broad lines. The two low velocity lines for the clumps
appear at $\approx$ - 0.7 mm/sec and $\approx$ 1.0 mm/sec; only the latter
can be seen.

10. CONCLUSION

The rare gas matrix isolation method in combination
with the Mössbauer effect is extremely useful in determining
nuclear parameters. This technique allows the implantation
and observation of neutral iron atoms and consequently the
isomer shift is determined for a known electronic configuration. An improved estimate of the ratio $\Delta R/R$ can be made
with this datum. The absence of large crystal field effects
simplifies greatly the calculation of the electric field
gradient at the Fe^{57} nucleus. This leads to a more reliable
value for the first excited state quadrupole moment.

There does not appear to be any compelling reason
which would limit the extension of the technique to other
Mössbauer nuclei. Most of those with gamma ray transition
energies below 40 keV appear to be suitable candidates for
absorber experiments. In all cases, it can be expected that
the resonant effects will be smaller than for Fe^{57}; however,
with the use of a variety of techniques (thicker absorbers,
cooled sources, higher count rates, etc.) it would appear
that success is assured.

The range of doable nuclei can be extended further with
the use of implanted sources and possibly, scattering
techniques. Micklitz [17] has attempted Cs^{133} in xenon using
a 5 Ci source and a cooled scatterer. The results appear
to show that, even for $f \approx 10^{-5}$, a small effect might be
observable.

Other experiments that can be performed with this
technique are more chemical in nature. Basic information
could be obtained on radicals and molecules such as FeO,
FeO_2, FeN, iron dihalides, etc.; they could be formed by
mixing a small amount of the active gas with the rare gas
before the implantation process, as was done in the iodine
experiment [3]. Transition metal molecules, e.g. Fe-Ni,
if they exist, could be created by using two furnaces to
deposit these materials simultaneously into the matrix. It
would be possible, of course, to extend further the scope of
these projected investigations with the substitution of
other Mössbauer nuclei for Fe^{57}.

The conclusion to be drawn from the foregoing discussion is that in future the use of the matrix isolation
technique will not be limited primarily to optical investigations but rather will be extended greatly by the application of Mössbauer spectroscopy.

ACKNOWLEDGMENTS

We would like to thank Dr. H. Broida and in particular
D. Mann, both of the optical spectroscopy group in our
department, for invaluable help and advice in the construction and operation of the matrix isolation apparatus.
We acknowledge very helpful and interesting discussions
with Drs. V. Jaccarino and H. Micklitz on the Fe^{57} results.
R. Marcus wrote the programs for the computer drawn
diagrams.

REFERENCES

1. For a recent review of this technique and the data ob-
    tained see W. Weltner, Advan. High Temp. Chem. 2, 85
    (1969), and references contained therein.
2. K. Mahesh, J. Phys. Soc. Japan 28, 818 (1970), and
    references contained therein.
3. S. Bukshpan, C. Goldstein and T. Sonnino, J. Chem. Phys.
    49, 5477 (1968).

4. P. H. Barrett and T. K. McNab, Phys. Rev. Letters, 25, 1601 (1970).
5. Encyclopedia of X-rays and Gamma Rays, Ed. G. L. Clark (Reinhold, New York, 1963), p. 9.
6. G. K. Wertheim, in: Mössbauer Effect Methodology, Vol. 4, Ed. I. J. Gruverman (Plenum Press, New York, 1969), p. 159.
7. L. R. Walker, G. K. Wertheim and V. Jaccarino, Phys. Rev. Letters, 6, 60 (1961).
8. J. Danon, in: Chemical Applications of Mössbauer Spectroscopy, Eds. V. I. Goldanskii and R. H. Herber (Academic Press, New York, 1968) p. 160, and references contained therein.
9. D. A. Shirley, Rev. Mod. Phys. 36, Pt. 2, 339 (1964).
10. H. Micklitz, to be published.
11. D. M. Mann and H. P. Broida, J. Chem. Phys., to be published.
12. For data on the solid rare gases see G. L. Pollack, Rev. Mod. Phys. 36, 748 (1964).
13. R. Ingalls, Phys. Rev. 133, A787 (1964).
14. A. J. Freeman and R. E. Watson, Phys. Rev. 131, 2566 (1963).
15. C. E. Johnson, W. Marshall and G. J. Perlow, Phys. Rev. 126, 1503 (1962) and references contained therein.
16. G. W. Robinson, J. Mol. Spectroscopy, 6, 58 (1961).
17. H. Micklitz, private communication.

CONVERSION ELECTRON MOSSBAUER SPECTROSCOPY

JON J. SPIJKERMAN

SCIENTIFIC RESEARCH CORPORATION

ALVA, OKLAHOMA   73717

Although most ^{57}Fe Mössbauer spectra are obtained with transmission geometry, several reflection (backscattering) techniques have been recently published. The scattering technique eliminates the need of sample preparation and provides a method for nondestructive testing of surfaces. The initial scattering experiments made use of the re-emitted 14.4 keV gamma radiation. However, due to the internal conversion process, this method is only 10% efficient, and the radiation is converted into 6.3 keV X-rays and conversion electrons. The atomic absorption of the gamma and X-ray radiation limits the penetration depth to 0.0005 inches. The 7.3 and 5.5 keV conversion electrons have a very limited range, less than 3000 Å in iron. The detection of X-rays thus primarily provides a bulk sample analysis, while the conversion electrons provide information on thin surface films (50 to 3000 Å), with applications in surface corrosion, surface absorption, catalysis, surface stresses, etc.

The backscattering technique requires the design of special detectors, shielding and source-sample-detector geometry. The design of these detectors is described, and the merits of the various detection techniques compared. Applications of the conversion electron detector are discussed.

INTRODUCTION

Mössbauer measurements have been made mainly in a
transmission geometry due to the simplicity of the experimental arrangement and large counting rates. Backscattering experiments were already discussed at the Allerton
Park conference [1], but have only been used for special
purposes, such as the study of the recoilless fraction of
the scattered radiation [2], observation of Rayleigh scattering [3,4], for nuclear anomalous dispersion in iron [5],
and polarization of Bragg-reflected gamma radiation [6].
Debrunner and Frauenfelder [7] demonstrated that for very
low recoilless fractions the backscattering technique gives
better results. In many cases a transmission measurement
is not possible, particularly in nondestructive testing and
surface analysis. These applications led to the development of backscattering techniques, which will increase the
use of Mössbauer spectroscopy as an analytical tool.

BACKSCATTERING TECHNIQUES FOR ^{57}Fe

The most direct method for observing the Mössbauer effect in backscattering geometry is to detect the re-emitted
14.4 keV radiation [8,9,10]. This method is inefficient
due to the large internal conversion of the 14.4 keV re-
emitted radiation. The background radiation is fairly low,
so that with good geometry and the use of Uranium metal for
shielding a 130% resonance effect can be obtained [10].
The resolution and efficiency of the detector are very important. The two inch diameter Krypton-filled counter is
normally used, but the recently developed high resolution
solid state detectors would give better results.

A second method for backscattering measurements is to
detect the 6.3 keV conversion X-rays. Hershkowitz and Walker [11] counted the conversion X-rays in a forward scattering geometry, and obtained good resonance effects from
very thin films of iron. The X-ray fluorescence from the
122 keV precurser in the source is a major contribution to
the background count. The efficiency for backscattering

using conversion X-rays is about three times larger than
for the 14.4 keV gamma ray detection, but the noise level
is much higher, so that both methods give comparable results [10,12].   Shielding is essential, and a 0.5 inch diameter Argon-filled proportional counter has a 65% efficiency and good resolution for the 6.3 keV X-rays.   The
source should be filtered by a 0.1 inch thick plexiglass or
0.005 inch aluminum foil attenuator, usually placed over
the absorber.   The geometry required for backscattering reduces the solid angle required for high efficiency.   Chow,
et al [13] developed a toroidial counter for higher geometrical efficiency.   A 2π steradians solid angle can be
achieved by passing the gamma radiation through the counter
[14].   In this detector the efficiency for the 14.4 keV radiation is an order of magnitude lower than the 6.3 keV detection efficiency, and the energy resolution sufficient to
differentiate between the 14.4 and 6.3 keV radiation.

The effective penetration depth in the sample is approximately 0.0005 inch (1.3×10^{-3} mm).   The backscattering
amplitude as a function of penetration depth into the sample is shown in figure 1, from the calculations of [12].

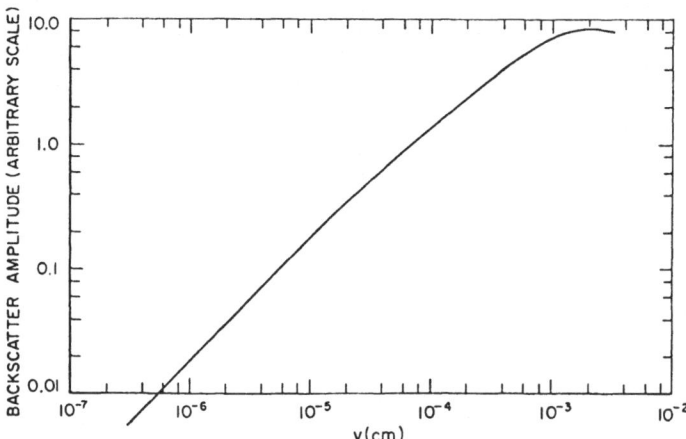

Fig. 1.   Number of 6.3 keV internal conversion X-rays from
Fe as a function of the penetration into the sample for
backscattering geometry.

To check their results, backscattering spectra shown in figure 2, were taken of a stainless steel foil covered with
several layers of thin metalic iron foils [14]. These results checked within a few per cent of the calculated value.
The effective penetration depth for the 14.4 keV radiation
is nearly the same as for the conversion X-ray radiation,
since the atomic absorption edge for iron is at 7.1 keV,
and the mass attenuation coefficients are nearly equal.
The linewidth of the conversion X-ray Mössbauer spectra are
narrower than those for the 14.4 keV gamma radiation due to
self-absorption of the resonant radiation in the sample after the Mössbauer effect takes place.

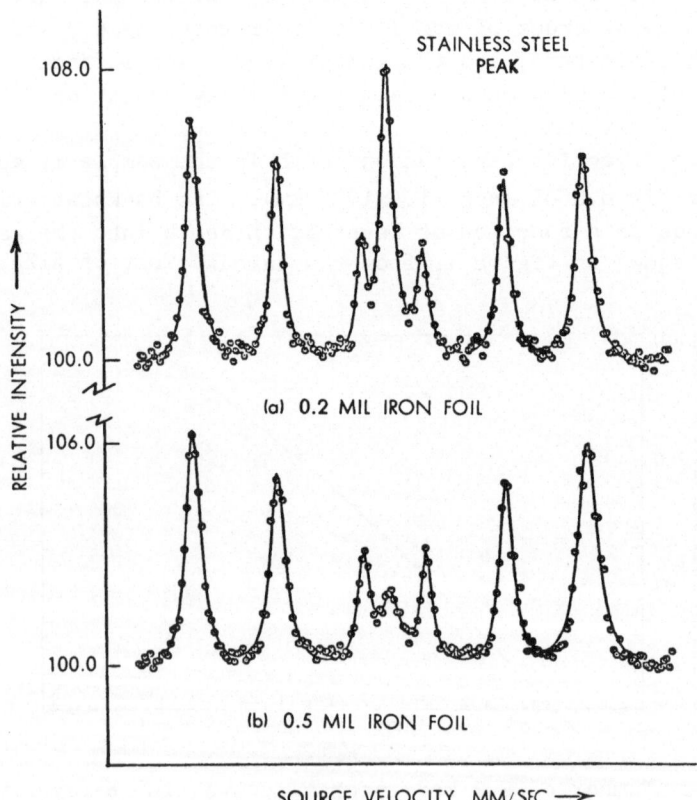

Fig. 2. X-ray detection Mössbauer scattering spectra for
iron foil on stainless steel foil.

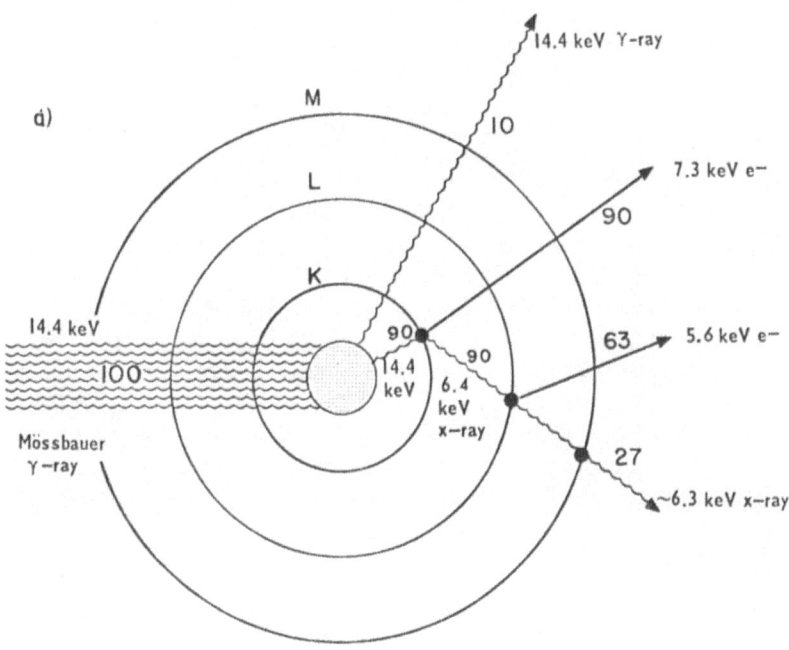

Fig. 3. a) Schematic of internal conversion process for the 14.4 keV transition of ^{57}Co. b) Energy degradation process for a source, absorber and detector system in a conventional Mössbauer detection system.

The third method used in backscattering techniques is
the detection of conversion electrons. This only applies
to very thin surfaces, between 50 to 3000 $\overset{o}{A}$, and requires
the absorber to be inside the detector. This method has the
advantage that very small samples (10 µg/cm^2) can be used,
and self-absorption broadening is absent which results in
a spectrum with very narrow lines predicted by the thin absorber limit. The principle of operation is based upon the
detection of internal conversion electrons emitted by reso-
nance-excited nuclei in the detector. The background in
this type of detector arises from secondary photoelectrons.
The process of decay of the 14.4 keV excited state in ^{57}Fe
is shown schematically in figure 3. For every 100 nuclei
in the 14.4 keV excited state ten nuclei decay by emitting
the 14.4 keV gamma ray, and the other 90 nuclei decay by
the internal conversion process (e.g. the internal conversion coefficient is 9). This means that a K shell electron
is emitted with approximately 7.3 keV energy. An electron
from the L shell enters the K shell and emits the characteristic 6.45 keV X-ray. Of these 90 - 6.4 keV X-rays, approximately 63 cause 5.6 keV electrons to be emitted by the
Auger effect, and the remaining 27 - 6.4 keV X-rays leave
the atom. M to L shell cascade produces less than 1 keV
X-rays. Internal conversion with the L shell occurs, but
with significantly less probability. Because of the fact
that out of 90 - 7.3 keV conversion electrons emitted there
are simultaneously 63 - 5.6 keV Auger electrons, the probability for electron detection is enhanced.

The efficiency and background noise of the electron
detector are determined by the processes which produce electrons and the probability of these electrons escaping from
the absorber to be detected. For the radiation in the energy range of interest, the case of a 57Co source, the only
significant contributions to the electron production are
Compton, photoelectric and resonant interactions. The
cross sections in barns are given in table I, from [15].
The resonance cross section is 2.38 x 10^6 barn, and the
major contribution to the background is the 8,450 barn
cross section of the photoelectric effect due to the 14.4
keV gamma radiation. The efficiency is limited by two

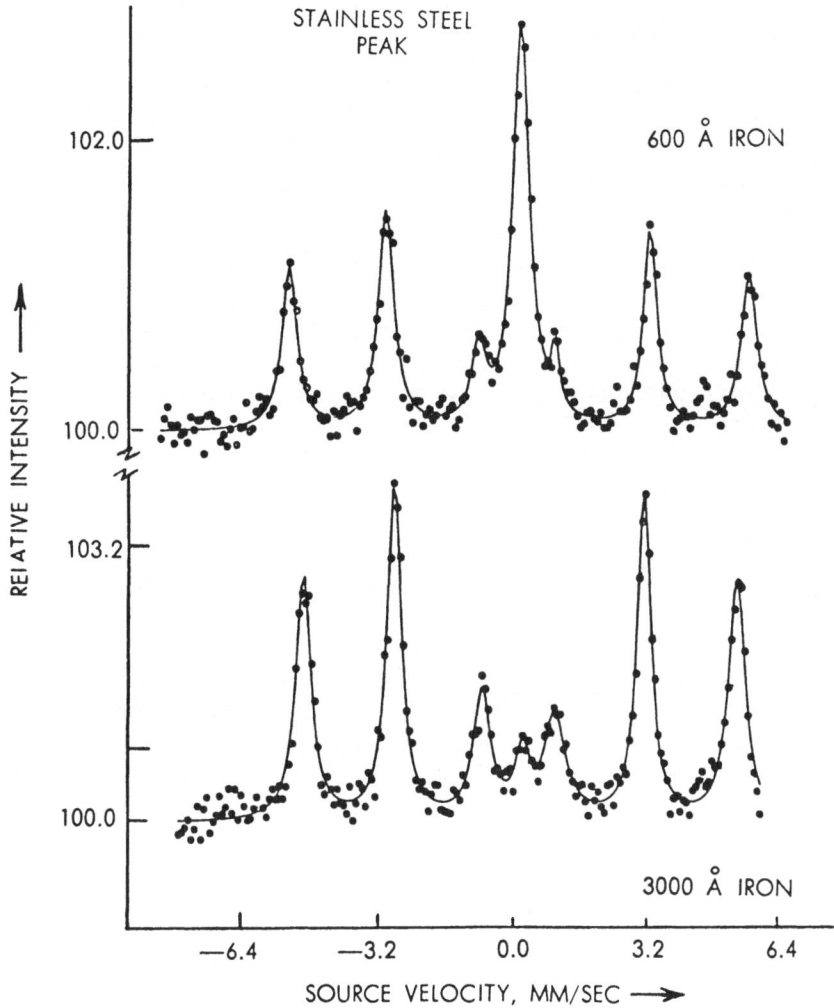

Fig. 4. Conversion-electron detection Mössbauer back-scattering spectra for vacuum-deposited iron on stainless steel foil.

essentially invariant factors, and the surface composition.

Table I.   Cross sections in barns for iron.

gamma ray	14.4	122	136
photoelectric	8450.0	12.2	8.5
Compton	16.4	14.2	14.5

The internal conversion coefficient places a limit of 90%
on the maximum efficiency, and the geometrical efficiency
of 0.5 places this combined limit at 45%. The primary
efficiency limit is the attenuation of the internal conversion electrons and accompanying Auger electrons in the sample material. The efficiency can be increased by changing
the geometry from 90° incident radiation to an almost parallel incidence with the sample surface.

The penetration depth for backscattering using electrons is below 3000 Å. Backscattering spectra [14] of a
stainless steel foil with 600 and 3000 Å metalic iron evaporated on the surface are shown in figure 4. The detection limit for iron metal would be about 50 Å, and with
enriched ^{57}Fe, it would be possible to detect a monolayer.
Due to polarization of the absorber, the intensity of the
2 and 5 lines in the iron spectrum are enhanced, indicating
that the magnetic field is parallel to the surface.

CONVERSION ELECTRON DETECTOR

A typical conversion electron counter, using a Helium-
6% CH_4 flow gas, is shown in figure 5. The counter has a
0.5 inch aperture, and is 0.25 inch thick, with two 1 mil
anode wires spaced for optimum efficiency and resolution,
and operates at 1200 volts. To check the detection uniformity of the counter, a lead-and-aluminum shield with a
0.05 inch-diameter pinhole restricted the radiation to the
desired area of the window. The pulse height spectra obtained at all points of the window were identical in terms

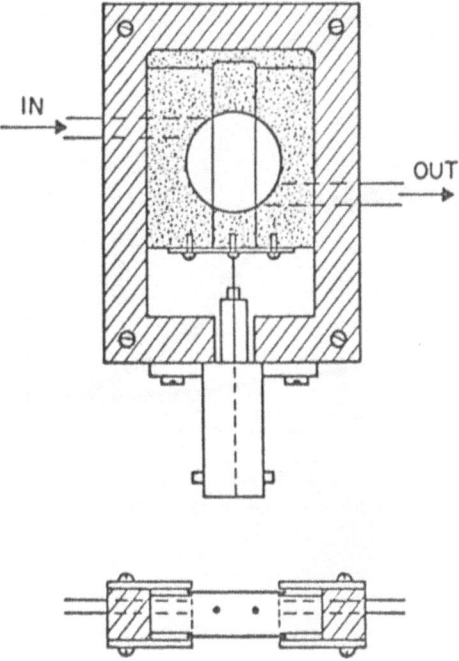

Fig. 5.   Diagram of detector for conversion electrons.

of resolution, counting efficiency and gas multiplication.
The resolution of the counter was further checked by placing
a ^{55}Fe source inside the counter, which gave the typical
endpoint spectrum at 5.5 keV, compared to the endpoint of
7.3 keV for the ^{57}Co source. To evaluate the noise contribution as a function of energy, a commercial 0.5 mil 50%
enriched in ^{57}Fe stainless steel foil was placed inside the
counter. The pulse height spectrum of a moving ^{57}Co source
in a stainless steel matrix with this detector is shown in
figure 6a, and with the source at rest in figure 6b. The
Mössbauer spectrum for this detector is shown in figure 7,
with a 325% effect. The optimum threshold for the energy
discrimination is above 2 keV. The maximum signal/noise
is obtained at a threshold setting of 4 keV, but the counting rate is too low to be practical. In the case of very
thin films on low Z substates, both windows of the counter
can be used for mounting the sample, dubling the efficiency.

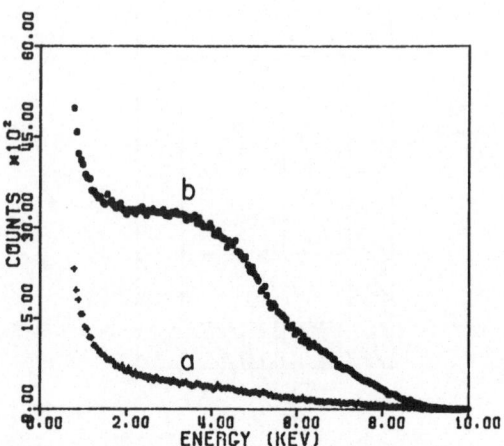

Fig. 6. Energy distribution of photoelectrons and conversion electrons obtained by source-detector off resonance (a)
and in resonance (b).

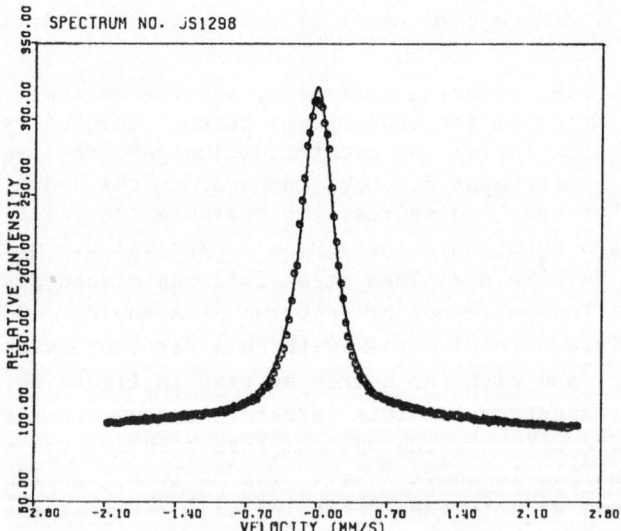

Fig. 7. Mössbauer spectrum of the resonance detector with a
0.5 mil 50 percent enriched stainless steel foil and moving
^{57}Co in stainless steel source.

APPLICATIONS

The various techniques for backscattering Mössbauer spectroscopy offer a surface analyzer over the range of 50 Å to 10^5 Å (0.5 mils) of iron bearing material nondestructively. By using chemically tuned sources, ^{57}Co in various source matrices, it is possible to use a stationary source-absorber method, and eliminate the need for a velocity scanning spectrometer, hence greatly increasing the data accumulation rate. Determination of surface compound formation [12] gave good results on corrosion product formation on steel plates. The Mössbauer spectral parameters of the iron oxides and oxyhydrides are characteristic for each material, and can be used for identification.
In electrochemistry, this technique would be extremely important for the study of passification of surfaces, and with thin Berylium windows transparent to the conversion electrons, the effect can be studied in situ. Although Mössbauer measurements have been made on catalytic interactions [16,17,18], the thick matrix is a problem for X- and gamma ray backscattering. Conversion electron detection would make it possible to work with very thin surface layers, particularly in the case of enriched materials.

The study of magnetic structure as a function of depth has many applications in bulk materials, two dimensional magnets (thin films), transitional magnetic properties in small particles, and crystal structure using the various faces of the crystal rather than prepared thin slices. Technical applications would be primarily in the steel industry for the measurement of phases in alloys, retained Austenite [13], surface hardening by cementation and nitration or cold working, and for the analysis of the interface in coatings. Since the internal magnetic field and isomer shift depends on stress and strain in steels, this detector could be developed into a surface stress-strain indicator.

REFERENCES

1. H. Frauenfelder, <u>The Mossbauer Effect</u>, W.A Benjamin, Inc., New York, 1963.
2. C. Tzara and R. Barloutaud, Phys. Rev. Lett. <u>4</u>,405 (1960)
3. P. J. Black, et al., Proc. Roy. Soc. (London) <u>A270</u>, 186 (1962)
4. P. J. Black and P. B. Moon, Nature <u>188</u>, 481 (1960)
5. S. Bernstein and E. C. Campbell, Phys. Rev. <u>132</u>, 1625 (1963)
6. J. Olsen, Nucl. Instr. and Methods, <u>70</u>, 109 (1969)
7. P. Debrunner and H. Frauenfelder, Tech. Rep. Ser. Int. At. Energy Agency (Vienna) <u>50</u>, 58 (1966)
8. A. N. Artmev et al., JETP <u>25</u>,768 (1967)
9. J. K. Major, Nucl. Phys. <u>33</u>, 323 (1962)
10. R. N. Ord, Appl. Phys. Letters, <u>15</u>, 279 (1969)
11. N. Herskowitz and J. C. Walker, Nucl Instr. and Methods, <u>53</u>, 273 (1967)
12. J. H. Terrell and J. J. Spijkerman, Appl. Phys. Letters, <u>13</u>, 1 (1968)
13. H. K. Chow et al.,"Mossbauer Effect Spectrometry for Analysis of Iron Compounds", AEC Rept. NSEC-4023-1 TID-4500, Contract No. At-(30-1)-4023 (1969)
14. K. R. Swanson and J. J. Spijkerman, Journal Appl. Phys. <u>41</u>, 3155 (1970)
15. J. J. Spijkerman, J. C. Travis, P. A. Pella and J. R. DeVoe, NBS Technical Note 541, U. S. Government Printing Office, Washington, D. C. 20402 (1971)
16. M. C. Hobson, Jr., J. Electrochem Soc. <u>115</u>, 175C (1968)
17. T. Yoshioka, J. Koezuka and I. Toyoshima, J. Catalysis, <u>14</u>, 281 (1969)
18. M. C. Hobson, Jr. and H. M. Gager, 4th Intern. Congr. Catalysis, Moscow 1968, Prepr. No 48.

Paramagnetic Hyperfine Structure in Iron

Complexes with Effective Spin S=1/2[*]

W. T. Oosterhuis

Physics Dept., Carnegie-Mellon University

Pittsburgh, Pennsylvania 15213

I. INTRODUCTION

Several paramagnetic iron compounds with spin S=1/2 are known to exist[1-4] in contrast to the more common high spin iron materials which have spin S=5/2 even though there are five valence electrons associated with the iron ion in each case. These low spin iron compounds can exhibit paramagnetic hyperfine structure (PHS) in Mössbauer spectra under the proper conditions[5] and the shape of the spectrum can provide information about the spin and charge distribution even in randomly oriented absorbers. The low spin situation arises when the energy separation 10Dq between the t_{2g} (or d_ϵ) and e_g (or d_γ) states becomes larger than the repulsion between electrons in the same orbital. Then (assuming octahedral coordination) it is energetically favorable for the electrons in the e_g states to flip their spins and occupy the t_{2g} orbitals leaving fewer unpaired electrons - hence the "low spin" configuration. We will be discussing those iron materials which have only a single unpaired electron per atom and therefore a net spin of one half.

A model which works well in describing the electronic properties of these spin one half ions is to consider the unpaired electron to be in a molecular orbital

[*] Supported in part by the National Science Foundation and the Office of Naval Research

$$\psi = N(\psi_{Fe} + A_{\pi}\psi_L)$$

where ψ_{Fe} is the t_{2g} or e_g state of the Fe ion and ψ_L is a linear combination of ligand orbitals constructed so as to have the same symmetry as ψ_{Fe}. N^2 is roughly the fraction of time an electron in such an orbital spends on the iron ion. One can form a basis set of molecular orbitals and consider these states to be acted on by low symmetry crystal fields and the spin orbit interaction. This electronic Hamiltonian usually has the form

$$\mathcal{H}_e = Az^2 + Bx^2 + Cy^2 + \lambda\vec{L}\cdot\vec{S} \quad .$$

These interactions will lift the orbital degeneracy and leave a ground state Kramers doublet well separated from the excited states (a few hundred cm^{-1}) as in Fig. 1. Since the experiments of interest here are done at low temperatures only the ground state doublet will be occupied. {A + B + C = 0 by Laplace's Eq.}

In the case of Fe(III) there are five electrons distributed among the three t_{2g} orbitals (t_{2g}^5 configuration) and it becomes useful to think in terms of a hole in the filled t_{2g}^6 subshell so that we have a "one electron"

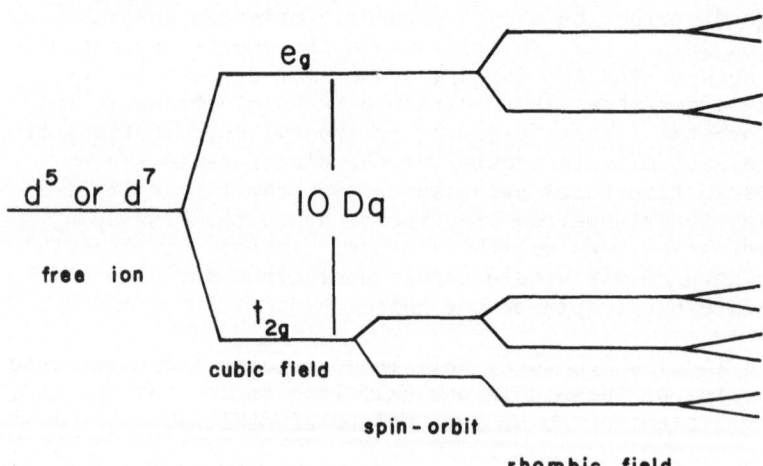

Fig. 1. One electron energy level scheme for 3d states. The repulsive electron-electron energy is not shown here.

problem. When $\mathcal{H}_e$ is diagonalized we will find states

$$\psi_+ = a|xy>\alpha + b|yz>\beta + ic|xz>\beta$$
$$\psi_- = a|xy>\beta - b|yz>\alpha + ic|xz>\alpha$$

as the members of the ground state Kramers doublet. The application of a magnetic field will lift the Kramers degeneracy and mix ψ_+ and ψ_-.

The occupation numbers of the hole a, b, c are determined by the low symmetry crystal field and the spin orbit interaction.[6] By finding a and b ($c=\sqrt{1-a^2-b^2}$) from the Mössbauer data, one can work backwards to find A/λ and B/λ. λ can be estimated from the temperature dependence of the susceptibility or the quadrupole splitting.

We will also consider the case of a low spin $3d^7$ configuration ($t_{2g}^6 e_g$) in which there is an unpaired electron (or hole) in $a|3z^2-r^2>$ orbital in the particular examples[7] considered here.

$$\psi_+ = |3z^2-r^2>\alpha$$
$$\psi_- = |3z^2-r^2>\beta$$

It is the goal of this discussion to provide an aid for the interpretation of magnetically split Mössbauer spectra of S=1/2 iron materials in terms of the electronic structure and environment although the method can be applied to any paramagnetic system where the Mössbauer effect can be observed. The spectra for spin one half iron materials are somewhat more complicated compared to the spin five halves situation since there are effects due to non-zero orbital angular momentum.

Only those paramagnetic systems in the limiting cases of extremely long ($\tau>>10^{-7}$ sec) or extremely short ($\tau<<10^{-7}$ sec) electron spin relaxation times will be considered here. The "static" case has been discussed in several papers[4,6,8-11]. The dynamic relaxation effects for intermediate relaxation times have been discussed by Wickman[8], Blume and Tjon,[12] and Gabriel.[13] The rapid relaxation case where one sees paramagnetic hyperfine structure by creating a large spin polarization has been reported by Obenshain et al,[14] Johnson[15] and Oosterhuis and Lang.[6]

II.   SLOW RELAXATION ($\tau \gg 10^{-7}$ sec).

Many examples of PHS in Mössbauer spectra have appeared in the literature in recent years.[6-11] The technique usually employed is to slow the electronic relaxation rate enough so that the Mössbauer nucleus can follow the fluctuating hyperfine field which in turn will split the nuclear levels. The relaxation rate is determined primarily by two mechanisms: spin-lattice relaxation which can be reduced by doing experiments at low temperature; and spin-spin relaxation which can be negated by diluting the para-magnetic spins in a diamagnetic medium. The effects of temperature and concentration have been documented by several workers[6,11] and the theory of the absorption spectrum as a function of relaxation rate has been worked out also.

One often sees that he is on the verge of the slow relaxation situation in concentrated salts because of the asymmetric broadening of the quadrupole split peaks[16] at low tem-perature. However, if one wants to see the full magneti-cally split pattern then the paramagnetic spins have to be separated in space which means that a suitable diluent has to be found. One can dilute in isomorphic diamagnetic lattices if they exist or in various solvents, preferably those which form glasses. We have tried water, DMF,[7] methanol, glycerol and frozen mixtures of these materials. Glycerol usually gives PHS but also gives very broad lines.[5] The methanol and DMF give fairly sharp lines but do not always dissolve the salt. In most biological molecules the iron is diluted naturally by the very large macro-molecules.

Once the slow relaxation situation is achieved ($\tau \gg 10^{-7}$ sec) magnetically split Mössbauer patterns can be observed due to the interaction between the electronic and nuclear moments. The shape or profile of the absorption spectrum is quite sensitive to the charge and spin dis-tribution of the electrons about the Mössbauer nucleus, and information about the local environment and the electronic state can be determined.

The Mössbauer spectrum reflects the several interactions included in the hyperfine Hamiltonian $\mathcal{H}$ which operates on the states $\psi_{\pm}$

$$\mathcal{H} = \beta_e \vec{H} \cdot (\vec{L} + 2\vec{S}) + 2g_n \beta_e \beta_n <r^{-3}> \{\vec{I} \cdot \vec{L} + 3(\vec{I} \cdot \hat{r})(\vec{S} \cdot \hat{r}) - (1+\kappa)\vec{I} \cdot \vec{S}\}$$

$$+ \frac{eQq}{4} \{I_z^{\,2} V_{xx} + I_y^{\,2} V_{yy} + I_z^{\,2} V_{zz}\} - g_n \beta_n \vec{H} \cdot \vec{I}$$

We can make a correspondence between the real Hamiltonian $\mathcal{H}$ operating on the real eigenstates $\psi_\pm$ and an effective spin Hamiltonian $\mathcal{H}_{eff}$ operating on eigenstates of the effective spin operator S_{eff}. $(S_{eff_z}|\pm> = \pm 1/2 |\pm>)$

$$\mathcal{H}_{eff} = \beta_e \vec{H} \cdot \tilde{g} \cdot <\vec{S}_{eff}> + \vec{I} \cdot \tilde{A} \cdot <\vec{S}_{eff}> + \vec{I} \cdot \tilde{q} \cdot \vec{I} - g_n \beta_n \vec{H} \cdot \vec{I} .$$

The first term is the interaction between the electronic moment and the applied magnetic field and is the dominant influence on the electronic moment for applied fields >50 Gauss. $<\vec{S}_{eff}>$ has the opposite directions for the two members of the Kramers doublet. The second term is the interaction between the nuclear and electronic moments with the electronic moment creating a field of several hundred kilogauss at the nuclear site whereas the nuclear moment creates a field of about 10 Gauss at the site of the electron. Thus there is a competition between the external field and the nuclear moment for the electronic moment which is won when the external field is larger than about 50 G. However, as far as the nucleus is concerned, the external field is negligible compared to the hyperfine field due to the electronic moment. The third term is the nuclear quadrupole interaction with the electric field gradient due to the non-symmetric electronic charge distribution. The fourth term is the direct effect of the external field on the nuclear moment which is negligible compared to the preceding terms for applied fields less than five kiloGauss.

The matrix elements of $\mathcal{H}_{eff}$ can be calculated within the basis set $|\pm>|I_z>$ and those matrix elements of $\mathcal{H}$ calculated in the basis $|\psi_\pm>|I_z>$. Then the $\tilde{g}$, $\tilde{A}$, and $\tilde{q}$ tensors can be identified in terms of the occupation numbers a, b, c for the t_{2g}^5 configuration and for the $t_{2g}^6 e_g$ case, both of which are listed in the appendix. The matrix has dimension $(2S_{eff} + 1)$ x $(2I+1)$ giving 8 for the nuclear excited state of ^{57}Fe $(I = 3/2)$ and four for the nuclear ground state $(I = 1/2)$ with $S_{eff} = 1/2$. The matrix elements for both excited and ground nuclear states are calculated with the applied field in some specified direction relative to the crystal axes. Each matrix is diagonalized to find the energies and eigenstates

and then the transition energies and probabilities are calculated for the Mössbauer transition. The Mössbauer spectrum
is then calculated with a line width Γ folded in assuming a
Lorentzian line shape. Γ arises due to the natural width of
source and absorber, plus relaxation and blackness effects.
This procedure can be repeated with Mössbauer spectra calculated for all orientations of the applied field relative
to the crystal axes corresponding to equal areas over the
unit sphere. The individual spectra for each field orientation are added together to give a simulation of the randomly
oriented powder absorbers. The amount of computation can be
reduced according to the symmetry of the Hamiltonian assumed.
Spectra are calculated assuming the γ beam to be unpolarized
and to propagate either parallel or perpendicular to the
applied magnetic field. Applied fields of a few hundred
gauss decouple the electronic spin from the nuclear moment
so that the electronic spin state depends only on $\vec{H}$. This
has the benefit of simplifying the Mössbauer spectrum by
eliminating the effects of small neighboring nuclear moments[17]
so that information about the symmetry of the electronic charge
and spin can be more easily obtained. The $\tilde{g}$, $\tilde{A}$, and $\tilde{q}$ tensors
depend mainly on the distribution of the electronic charge and
spin throughout the complex ion aside from a few factors such
as lattice EFG's, orbital reduction factors, and Sternheimer
factors which are not expected to change by much from molecule
to molecule. Thus, these tensors which determine the paramagnetic hyperfine structure in the Mössbauer spectrum depend
to a large extent only on a limited number of parameters a, b,
$<r^{-3}>N^2$, and possibly a quantity γ which is an angle describing the monoclinic crystal field.[18] The parameters a
and b describe the local symmetry of the electronic state in
terms of the 3 t_{2g} orbitals and can be used to find the ratios
$A/_\lambda$ and $B/_\lambda$. γ is an angle which is zero for orthorhombic
symmetry and $0<\gamma\leq45°$ in monoclinic symmetry. $<r^{-3}>N^2$ enters
into both $\tilde{A}$ and $\tilde{q}$ interactions and can be used as a factor
to adjust the scaling of the energy range of the Mössbauer
spectrum.

Examples of calculated spectra for randomly oriented
spin 1/2 Fe salts have been reported for a few cases of
varying symmetry of the electronic orbitals.[19] Those
calculations are extended here to cover a wider variety of

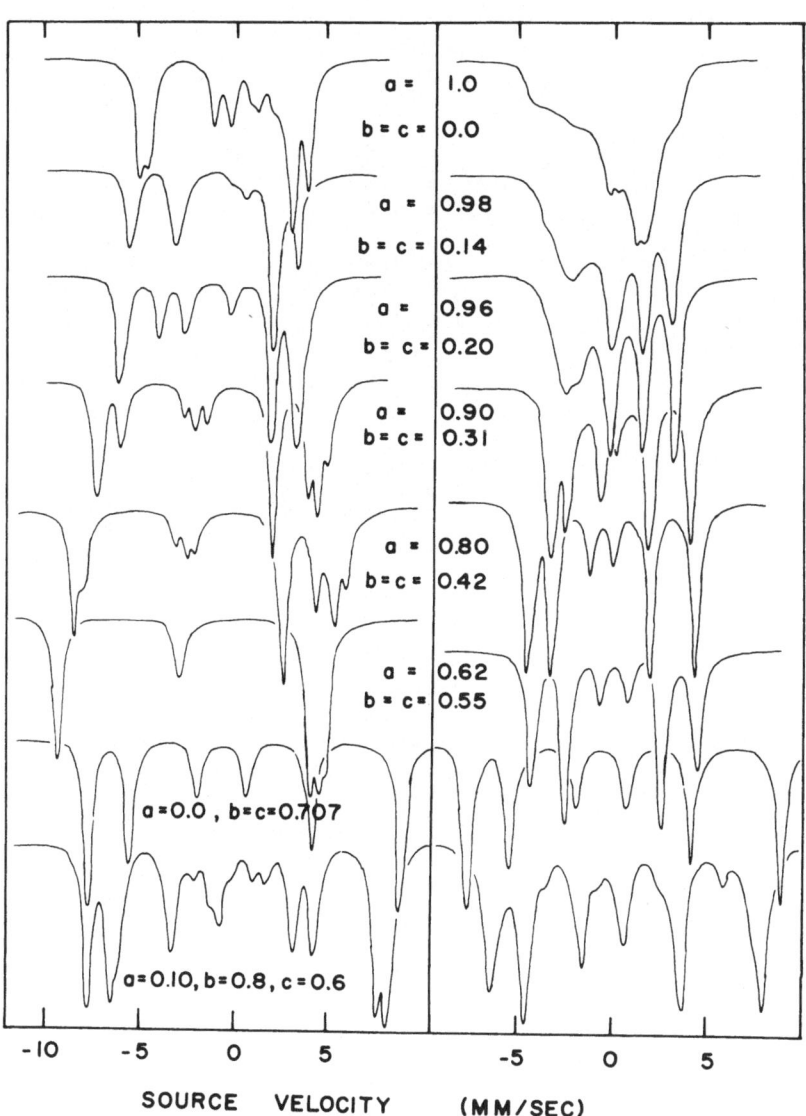

Fig. 2. Spectra calculated in the slow relaxation limit for randomly oriented absorbers in zero field (left) and with a field of 0.5 kilogauss oriented perpendicular to the gamma beam (right).

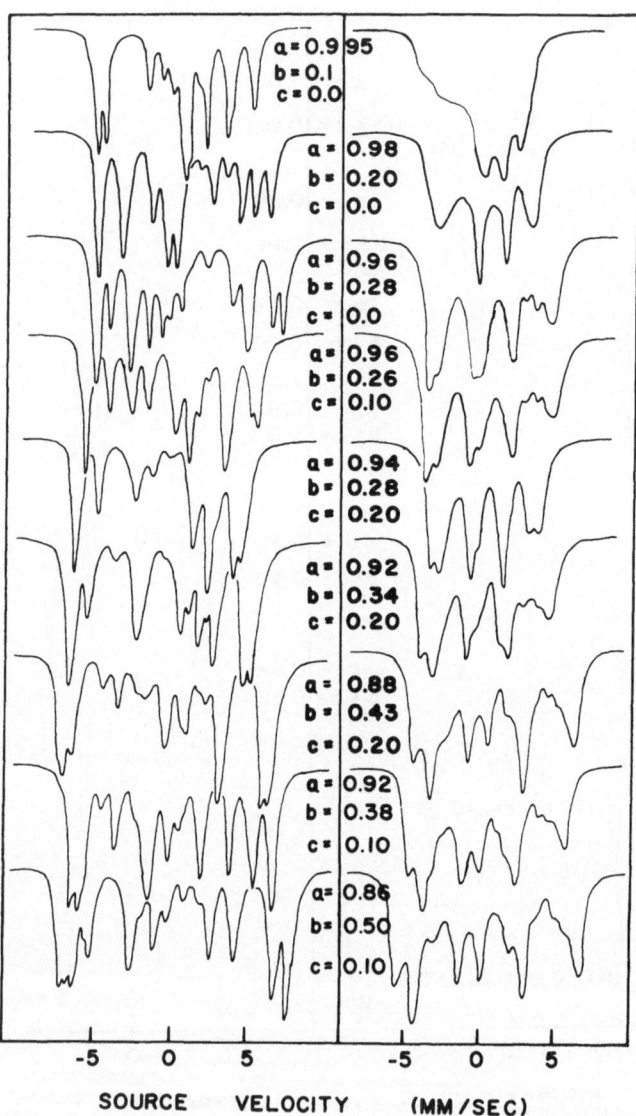

Fig. 3. More slow relaxation spectra calculated with zero applied filed (left) and 0.5 kilogauss transverse to the gamma beam for examples of orthorhombic symmetry of varying degrees.

electronic structures extending from a pure t_{2g} orbital
(a = 1.0, b = c = 0.0) to axial symmetry (a is varied,
b = c = $\sqrt{(1-a^2)/2}$) to rhombic (a and b are varied,
c = $\sqrt{1-a^2-b^2}$) to cubic (a = b = c = $1/\sqrt{3}$).

The parameters $<r^{-3}>N^2$, k,κ, 1-R which appear in the
expressions for $\tilde{g}$, $\tilde{A}$, and $\tilde{q}$, are chosen to have representative
values for Fe(III) materials[6] and are held constant through-
out the calculations. To fit individual spectra these parameters can be varied to optimize the agreement with the data.
It has been assumed that $<r^{-3}>N^2$ is the same for the magnetic and quadrupole interactions which is not true in
general.[20] However, it is a fair enough approximation for
the resolution one obtains in the experiments considered
here.

Calculated spectra for powder samples are presented for
the cases of zero applied field and a 500 G field perpendicular
to the γ-beam (Figs. 2-3). One can also use a 500 G field
parallel to the γ beam to obtain the same information as
the perpendicular field experiment, although the spectrum for
a given example will be different. However, it is experimentally
easier to use a field ⊥ to the γ-beam.

It is seen that a very small admixture of $|xz>$ or $|yz>$
($a^2 \neq 1.0$) into the ground state orbital $|xy>$ severely alters
the Mössbauer spectrum. On the other hand spectra for well
mixed orbitals do not change much with relatively large changes
in the coefficients (a, b, c). The quadrupole splitting becomes small as the asymmetry of the configuration becomes
less which fits in with the observed behavior of the quadrupole
splittings of these salts in zero field.[2] If the sign of
the electric field gradient were to be reversed, the spectrum
in an applied field would become its mirror image reflected
through the centroid. The change in the zero field spectrum
would be more complicated. The spectra calculated for powder
samples in an applied field are generally smeared out due to
the distribution of hyperfine fields seen by the nuclei. However, when axial symmetry exists, the Mössbauer transitions
are noticeably more sharp. This is because the two g-values ⊥
to the symmetry axis are equal and larger than the third.
Then when a field is applied, the electronic moment tends to
lie in the plane ⊥ to the symmetry axis because of the large

$g_\perp$ and produce a hyperfine field which has the same magnitude everywhere in the plane. Thus the Mössbauer transitions will have more nearly the same energy independent of orientation when axial symmetry exists and this gives sharp absorption lines.

If the complex ion has local monoclinic symmetry ($\gamma \neq 0$) this too can alter the Mössbauer spectrum because of the fact that the EFG and g-tensors become reoriented with respect to each other.[18] This is another variable that one can use to optimize the fit to experimental data, but which will not be discussed here.

Some experimental data for randomly oriented Fe(III) materials of varying symmetry are presented in Fig. 4. The effect of an applied magnetic field is shown and can be compared to the zero field spectrum. In each case, the applied field strength of a few hundred Gauss is enough to decouple the electronic spin from the nuclear moment and the resultant spectrum is characteristic of the electronic charge and spin distribution about the ^{57}Fe nucleus. The zero field spectra will sometimes indicate the presence of interactions between the electronic moment and the ligand nuclear moments. Such an interaction is indicated by the discrepancies between the calculated and experimental spectra even though the agreement in the presence of the applied field is relatively much better. This is seen for HiN_3 and $N_3Fe(CN)_5NO_2$ in Fig. 4.   A random field model   was used to simulate the effects of the N^{14} moment interacting with the electronic spin and this crude approximation did bring the calculated spectrum in closer agreement with the experiment. The correct method of treatment would include an additional term in the Hamiltonian with interaction between the electronic spin and the ligand nuclear moments. Such an interaction when measurable is a good indicator of the covalent sharing of the magnetic electrons. When very small fields are used, we see marked effects in the Mössbauer spectra of $K_3Fe_xCo_{1-x}(CN)_6$ and $Na_2Fe(CN)_5NH_3$ as shown in Fig. 5. The solid curves are calculated with no adjustable parameters. It appears that the coupling of the electronic moment to the nuclear moment is almost completely broken by applied fields of about 30 Gauss.

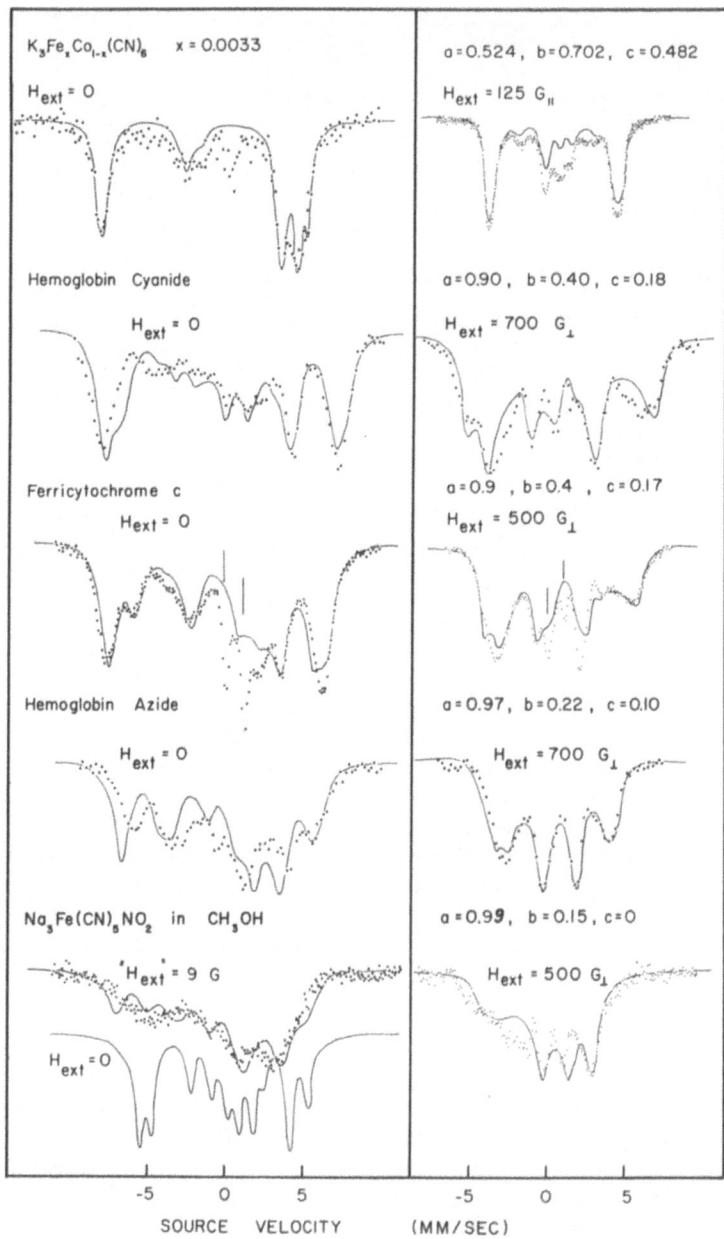

Fig. 4. Data of some dilute S=1/2 iron absorbers taken at 4.2K where the slow relaxation situation is approached. The solid curves are calculations like those of Figs. 2 and 3. Experimental data of heme proteins is that of Lang et al.[4,21]

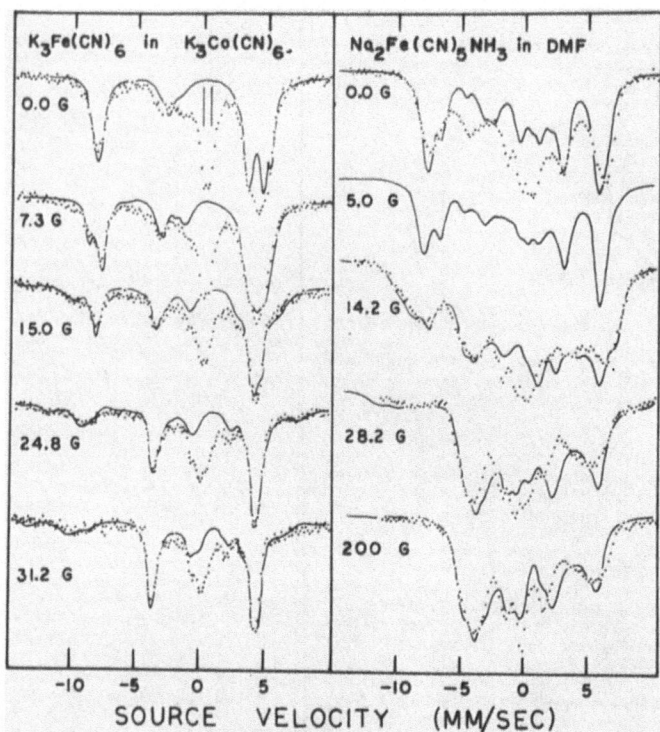

Fig. 5. Spectra of slowly relaxing paramagnets in very
small fields where the external field dominates the electronic moment at field strengths > 30G. Absorption near
0 mm/sec in $K_3Fe,Co(CN)_6$ data is due to some fast relaxing
iron sites. There may also be relaxation effects in the
$Na_2Fe(CN)_5NH_3$ which would broaden the resonances. The
solid curves represent calculations with no adjustable
parameters. (a) Before K_3; (b) After K_3.

The agreement between the calculated and experimental
spectra is optimized by varying a, b, and $N^2<r^{-3}>$. We see
that the agreement between the calculated and experimental
data is not very good in some places. However, it should
be kept in mind that the calculated spectra are obtained
by varying only 3 parameters for these powder spectra, but
in fact the iron nuclei see a distribution of fields due
to the random orientations of the crystals relative to the
applied field. There is also the very real possibility

that the static case is not realized here and there are
some relaxation effects in the data not taken into account in
the calculation.

The unusual case of a single unpaired electron in a   (7)
$|3z^2-r^2\rangle$ orbital about an Fe ion has recently been observed.
Mössbauer spectra of randomly oriented samples of reduced
sodium nitroprusside and HiNO identify the charge and spin
distribution of the electron by the shape of the spectrum in
small fields. The Mössbauer spectra characteristic of these
low spin d^7 cases are shown in Fig. 6 for zero and small
applied fields and the expressions for the $\tilde{A}$, $\tilde{g}$, and $\tilde{q}$
tensors are given in the appendix. One mystery that remains
is to understand the sign of the EFG - from the experimental
spectrum it appears as if there is a positively charged
electron in a $|3z^2-r^2\rangle$ orbital. The spectrum in a field
for a negative electron would be the mirror image
reflected through the 0 velocity axis due to the change in
sign of the EFG.

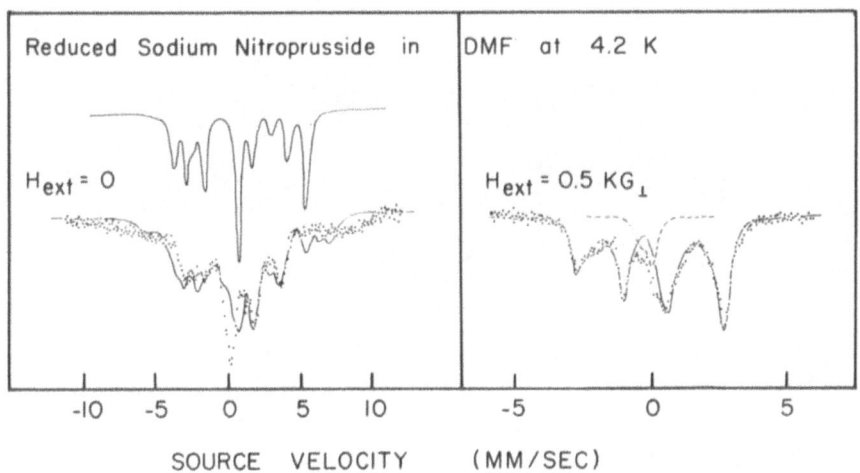

Fig. 6. Data of reduced sodium nitroprusside dissolved in
DMF (blue complex). Mössbauer data clearly indicate elec-
tronic state with $|3z^2-r^2\rangle$ symmetry. Data in a small field
were matched to the calculation by scaling the hyperfine
interaction. The zero field calculation with the same
parameters is shown as the isolated solid curve. When the
transferred hyperfine interaction with ^{14}N is taken into
account, the agreement between experiment and calculation
is greatly improved.

In the case of $Fe(CN)_5NO^{-2}$ (blue) in DMF there was an additional interaction between the magnetic electron and the nuclear spin of the N^{14} of the NO ligand. This transferred hyperfine interaction has the effect of altering the Mössbauer spectrum as shown in Fig. 6 . In order to calculate the effect of this additional interaction, an additional term was added to the Hamiltonian $\mathcal{H}_{eff} + \vec{I}_N \cdot B \cdot \vec{S}_{eff}$. However, the dimension of $\mathcal{H}_{eff}$ is then increased to $(2I_{Fe}+1)(2S_{eff}+1)$ x $(2I_N+1)$ as the basis set then becomes $|I_{z(Fe)}\rangle|S^{\pm}_{eff_z}\rangle|I_z(N)\rangle$. We see from the figure that the additional interaction improves the agreement between calculation and experiment but doesn't make it perfect. A small amount of impurity accounts for the extra absorption as noted by the dashed lines.

Other examples of slow relaxation PHS in biological molecules are found in References 22 and 23.

III. FAST RELAXATION CASE

In many cases where one might want to observe hyperfine interactions in a paramagnetic system, it is difficult to attain the slow relaxation rates needed for observing the effect. Also it is usually difficult to find a diamagnetic host for diluting the paramagnetic ion in question and if one does succeed in finding a workable diluent, then questions arise concerning the change in the paramagnetic ion due to the unnatural environment.

However, if the system is relaxing quickly then one can still use PHS to determine the electronic state in a concentrated salt by an alternative approach. If the relaxation rate $>> 10^7$ sec^{-1}, then the hyperfine field seen by the nucleus is due to the thermal average of the electronic moment.

The appropriate Hamiltonian is $\mathcal{H}_{eff}$

$$\mathcal{H}_{eff} = \vec{I} \cdot \overset{\approx}{A} \cdot \langle \vec{S}_{eff} \rangle_T + \vec{I} \cdot \overset{\approx}{q} \cdot \vec{I} - g_n \beta_n \vec{H} \cdot \vec{I}$$

$$\text{where} \quad <\vec{S}_{eff}>_T = \sum_{n=-}^{n=+} <n|\vec{S}_{eff}|n> \; e^{-E_n/kT} / \sum_{n=-}^{n=+} e^{-E_n/kT}$$

with E_n the energies of the two members of the Kramers doublet. $<\vec{S}_{eff_i}>_T$ is made large by applying a large external magnetic field to split the Kramers doublet so that the low energy member will be mostly occupied at sufficiently low temperatures. The two members of the doublet give matrix elements $<n|\vec{S}_{eff}|n>$ which cancel each other, but the fact that the two members are unequally populated leads to a non-zero $<S_{eff}>_T$ because of the Boltzmann average. Thus one tries to achieve a situation where H/T is as large as possible. Magnetic fields of about 30 KG at temperatures of 4°K are usually large enough to show significant PHS in Mössbauer spectra. Usually, the Mössbauer spectrum will be more complicated due to the direct effect of the applied field in such a situation.

The Hamiltonian matrices now operate only on the nuclear basis states $|I_z>$ since the electronic state fluctuates much faster than the nucleus can sense the individual states. The matrices for the excited and ground nuclear states are diagonalized as before and the Mössbauer spectra are calculated in the same way and the powder average is then made.

This technique which we call the "fast relaxation" method can be used in all paramagnetic materials since the applied field induces a moment even in integral spin systems. The slow relaxation method won't work in general for integral spin paramagnets since the ground state will be a singlet and (in the absence of an applied field) have no moment to couple to the nuclear moment. However, an applied field will induce a moment (if it is a paramagnet) and if it is large enough will show PHS.[6,14,15] Other advantages of the fast relaxation technique are that one doesn't have to bother about finding diluents for paramagnetic samples, and that one doesn't have to worry about changes in chemical structure brought about by the diluent.

Thus one needs a superconducting magnet to produce fields greater than 30 KG at 4°K or one can pump on liquid

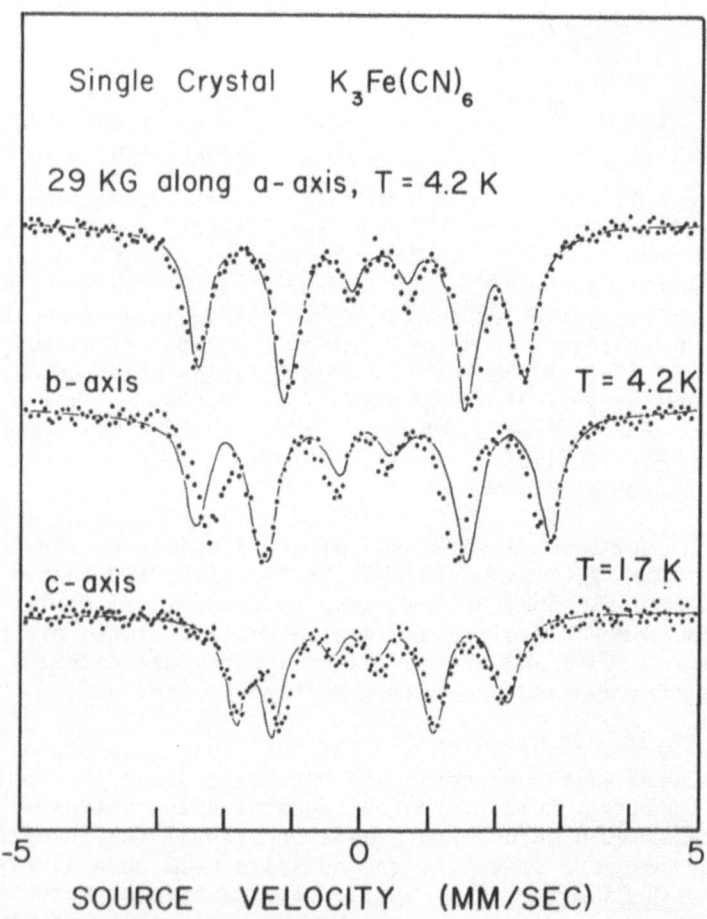

Fig. 7. Single crystal K₃Fe(CN)₆ in the fast relaxation regime with fields applied along the a, b, and c crystalline axes. Since the g-value in the c-direction is so small the magnetic splitting is also small and a temperature of 1.7°K was necessary to resolve the lines in 29 KG. The solid curves are calculated with the parameters used for K₃Fe,Co(CN)₆. The disagreement between calculation and experiment can be attributed to the difference between the concentrated and dilute salts.

He to get temperatures down to 1.5°K. However, with the recent development of He^4-He^3 dilution refrigerators(24) giving temperatures as low as 0.04°K one needs only an applied field of about 800 G to create the same $H/_T$ as the super-conducting magnet giving 30 KG at 1.5°K. Since the direct effect of the applied field will be negligible in these cases the spectra in saturated $<S_{eff}>_T$ will look just like the slow relaxation spectra in Figs. 2 and 3. One then avoids com-plicating the Mössbauer spectrum by the direct effect of a large external field.

Of course when one gets to such low temperatures, one will probably find most paramagnets becoming magnetically ordered. But this is just the case where you don't have to apply any external field at all! There is a spontaneous magnetization (or sublattice magnetization in the case of an antiferromagnet) to create the electronic moment that couples to the nuclear moment.

The technique of using large values of H/T can be used in single crystals to determine the principal values of the $\tilde{g}, \tilde{A}$ and $\tilde{q}$ tensors in the crystal. This has been done for single crystals of $K_3Fe(CN)_6$ as shown in Fig. 7. However, if one has only a powder specimen, then the values of these tensors can still be found because the Mössbauer spectrum will have a characteristic shape for a given electronic dis-tribution. Of course one can't determine the orientation of these tensors relative to the crystal axes in a powder. In the case of $Na_2Fe(CN)_5NH_3 \cdot H_2O$ we see how the spectrum changes with the applied field and the temperature in Fig. 8. One tends to expect that the determination of the electronic state (a, b, c) will be unique because of the many lines and low symmetry of the Mössbauer spectrum. Examples of Mössbauer spectra calculated with the same electronic model as used in the slow relaxation case (same values of $\tilde{g}, \tilde{A}, \tilde{q}$) are shown in Fig. 9 where a magnetic field of 30 KG is applied to powder samples in a direction perpendicular to the γ beam. Spectra calculated for fields parallel to the gamma beam are usually quite different in profile although the same information is obtained. Of course wide variations are obtained with varying H and T, but the examples presented here show representative behavior. Again we see that the resonance lines are narrow for axial symmetry $b=c=\sqrt{(1-a^2)/2}$.

Some experiments were done at 1.7°K in fields of 29KG as shown in Fig. 10. From the experimental data for

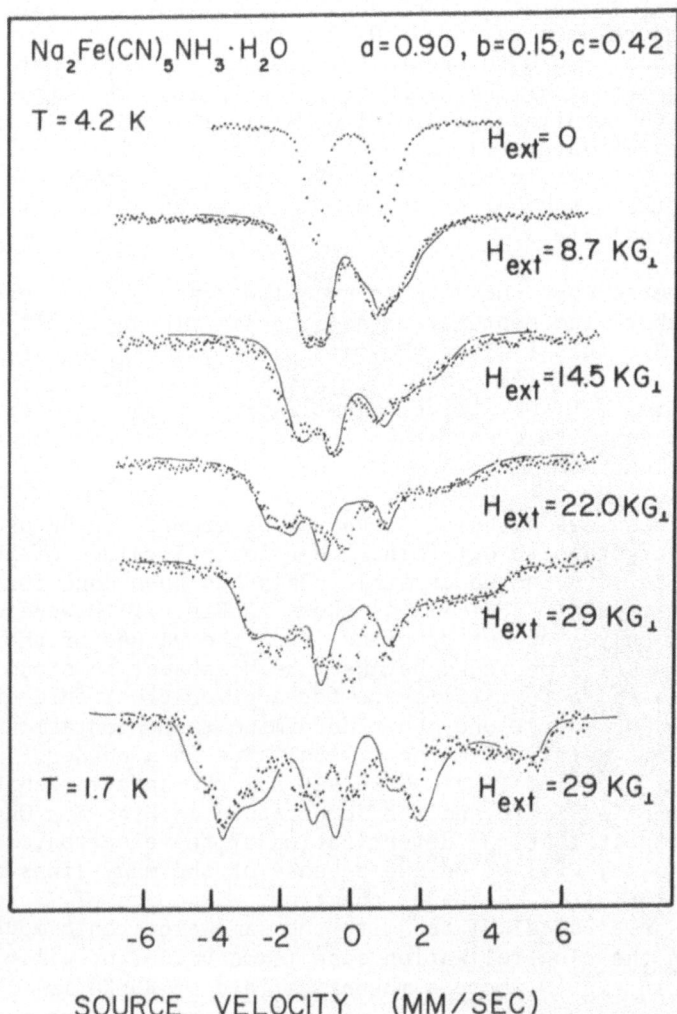

SOURCE   VELOCITY   (MM/SEC)

Fig. 8. The effect of the magnetic field strength and temperature on the spectrum of randomly oriented $Na_2Fe(CN)_5NH_3$ (concentrated salt) in the fast relaxation situation. The solid curves are calculated with a = 0.90, b = 0.40, c = 0.20 and the only variables are the applied field strength and temperature.

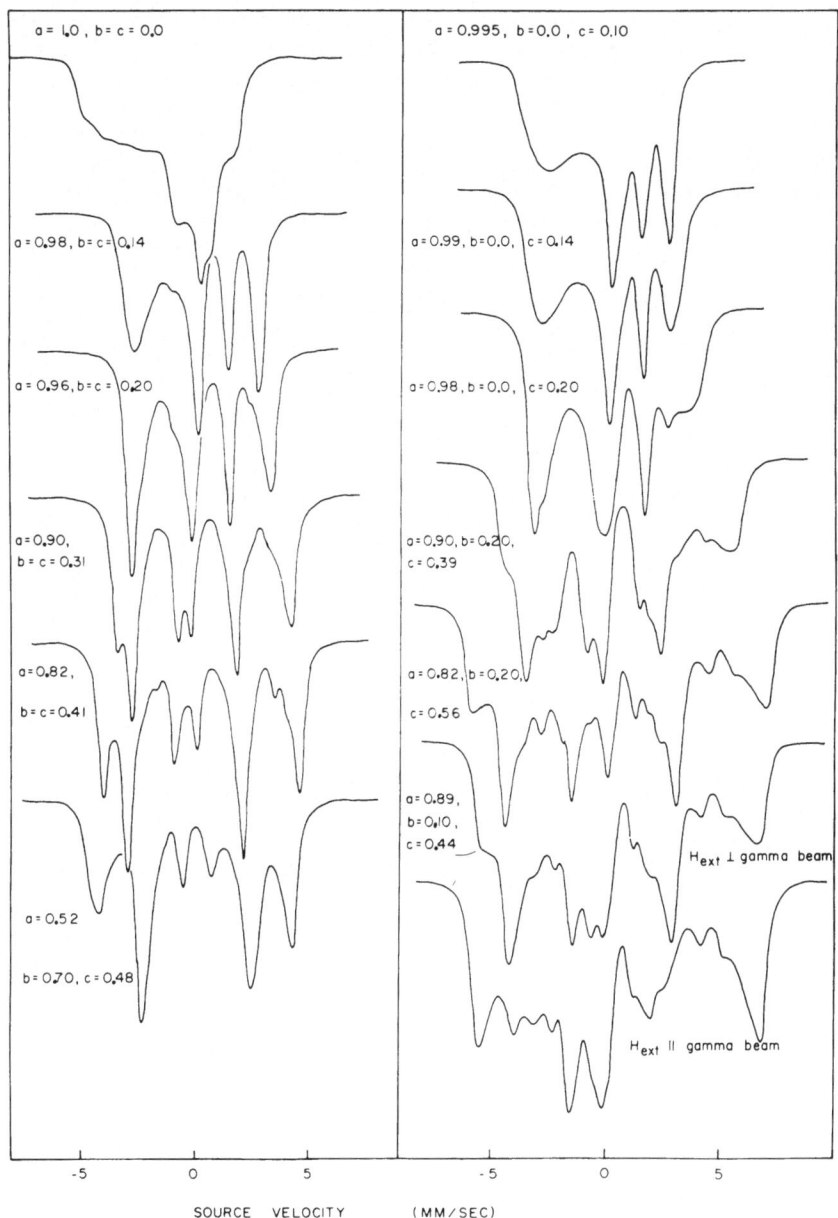

Fig. 9. Calculated spectra in 30 KG at 1.7°K for randomly oriented S = 1/2 absorbers. The effect of changing symmetry can be readily seen.

$Na_2Fe(CN)_5NH_3 \cdot H_2O$, $Na_3Fe(CN)_5NO_2$, and $Fe(das)_2Cl_2 \cdot BF_4$ we are
able to make an estimate of the parameters a, b. In the case
of $Fe(das)_2Cl_2 \cdot BF_4$ there is a very small hyperfine interaction
which indicates an extremely large delocalization of the
magnetic electrons to the ligands. This is supported by an
ESR measurement where the S=1/2 resonance was extremely
broadened due to excessive hyperfine interaction between the
electronic spin and the ligand nuclei.[25]

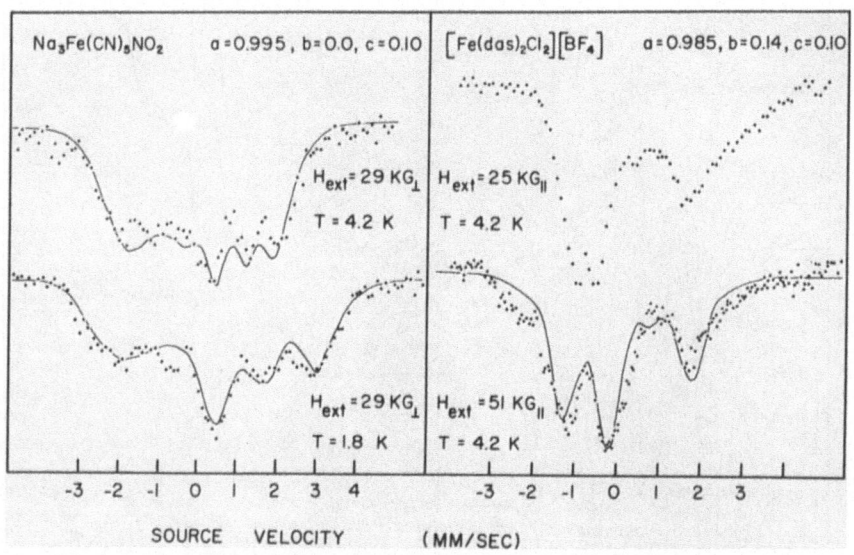

Fig. 10. Data for $Fe(das)_2 Cl_2 \cdot BF_4$ and $Na_3Fe(CN)_5NO_2$ in
large H/T. These are examples with extreme axial distortions as indicated by a, b, c.

IV.   CONCLUSIONS

Paramagnetic hyperfine structure can be observed in
Mössbauer spectra of spin 1/2 iron materials under conditions of very slow electronic relaxation ($\tau \gg 10^{-7}$sec),
or very fast relaxation ($\tau \ll 10^{-7}$sec) if a sufficiently
large spin polarization is created by magnetic ordering or
by application of an external magnetic field at low
temperature. The profile of a spectrum even for powder

absorbers can be directly related to the distribution of the
electronic charge and spin so that one has a measure of the
symmetry of the electronic state. The scaling of the spectrum
gives an approximate indication of the covalent sharing of
the magnetic electrons. The accuracies of the measured
values of the $\tilde{g}, \tilde{A}$, and $\tilde{q}$ tensors are probably within 10
or 15% of the true values.

Once one has determined the coefficients a_1, b_1, c_1 for
the electronic ground state ψ_1, then the characteristic
equations of H_e operating on ψ_1 can be used to determine
the ratios $A/_\lambda$, $B/_\lambda$, $E1/_\lambda$.

these ratios can in turn be used in matrix diagonal-
ization to determine the coefficients a_n, b_n, c_n and $F_n/_\lambda$ for
n=2,3. Then the susceptibility and electric field gradient
can be calculated as functions of λ/kT. By comparison of
the thermal behavior of the susceptibility or electric field
gradient (in powder samples) with the experimental data,
the spin orbit coupling λ can be deduced.

The rhombic crystal field plus spin orbit coupling model
works well in describing paramagnetic hyperfine structure of
many low spin iron salts. However, the more general case of
monoclinic symmetry can be done and applied to the same
low spin iron salts.[18]

This effect was not considered here but does allow for
another variable in optimizing the agreement between the
calculated and experimental spectra.

Zero field data exhibiting PHS may not be a reliable
indicator of the electronic state because there may be small
hyperfine interactions between the magnetic electrons and
neighboring nuclei. However, small fields of about 500
Gauss will decouple the electronic moments from all
nuclear moments and will give Mössbauer spectra characteristic
of the electronic state symmetry. If one then determines
the electronic state from the 500 G data and goes back to
the zero field data, information concerning the covalent
sharing of the magnetic electrons with neighboring nuclear
moments can be obtained by measuring the strength of the
transferred hyperfine interaction.

In slow relaxation cases, Mössbauer spectra of an iron material diluted in a diamagnetic host must be characterized as such because the host plays an important part in the electronic behavior of the iron ion. However, fast relaxation data for concentrated salts in large H/T situations are characteristic of the iron material itself. It is interesting that data for concentrated paramagnetic salts yield values which compare well with the numbers obtained from the same salts diluted in a foreign environment within the accuracy of the experiment.

When present, paramagnetic hyperfine interactions give more bumps in the Mössbauer spectrum and allow one to literally obtain a better grasp of the electronic structure.

ACKNOWLEDGEMENTS

The author is grateful for many helpful discussions with George Lang who also made his data on the haem proteins available. Thanks are also due to F. deS. Barros for several helpful comments and to E. A. Paez for help in some of the experimental work and analysis.

REFERENCES

1) L. A. Welo, Phil. Mag. 6, 481 (1928).
2) N. L. Costa, J. Danon, and R. M. Xavier, J. Phys. Chem. Solids 23, 1783 (1962).
3) B. N. Figgis, J. Chem. Soc. (A) 1966, 422.
4) G. Lang and W. Marshall, Proc. Phys. Soc. (London) 87, 3 (1966).
5) W. T. Oosterhuis, G. Lang, and S. DeBenedetti, Physics Letters 26A, 214 (1968).
6) W. T. Oosterhuis and George Lang, Phys. Rev. 178, 439 (1969). An alternative notation is used in Ref. 4. Appendix B provides a transformation.
7) W. T. Oosterhuis and George Lang, J. Chem. Phys. 50, 4381 (1969).
8) H. H. Wickman, Mössbauer Effect Methodology, Vol. 2 (Plenum Press 1966).
9) G. Lang, J. Appl. Physics 38, 915 (1967).

10) H. H. Wickman and G. K. Wertheim, Phys. Rev. 148, 211
    (1966).
11) L. E. Campbell and S. DeBenedetti, Phys. Rev. 167, 556
    (1968).
12) M. Blume and A. Tjon, Phys. Rev. 165, 446 (1968).
13) H. Gabriel, Phys. Stat. Solids 23, 195 (1967).
14) F. E. Obenshain, L. D. Roberts, C.F. Coleman and
    D. W. Forester, Phys. Rev. Letters 14, 365 (1965).
15) C. E. Johnson, Physics Letters 21, 491 (1966),
    C. E. Johnson, Proc. Phys. Soc. 92, 748 (1967).
16) H. H. Wickman, and A. M. Trozzolo, Phys. Rev. Letters 15,
    156 (1965).
17) G. Lang, T. Asakura and T. Yonetoni, Phys. Rev. Letters 24,
    981 (1970).
18) W. T. Oosterhuis, Phys. Rev. B3, 546 (1971).
19) George Lang and W. T. Oosterhuis, J. Chem. Phys. 51, 3608
    (1969).
20) R. E. Watson and A. J. Freeman, Magnetism, Vol. IIA, p. 167(Ed.
    G. Rado and H. Suhl, Academic Press, 1967).
21) G. Lang, D. Herbert, and T. Yonetoni, J. Chem. Phys. 49,
    944 (1968).
22) G, Lang, Quarterly Reviews of Biophysics 3, 1 (1970).
23) P. DeBrunner, Spectroscopic Approaches to Biomolecular
    Conformation, (Ed. D. W. Urrg, American Medical Assn.
    Press, Chicago, 1969).
24) G. M. Kalvius, T. E. Katila, O. V. Lounasmaa, Mössbauer
    Effect Methodology, Volume 5 (Plenum Press 1970).
25) D. Wood (Private Communication).

Appendix A. Spin Hamiltonian Tensors $\tilde{g}$, $\tilde{A}$, $\tilde{q}$ for t_{2g}^5 and t_{2g}^6 e_g Configurations

| t_{2g}^5 ($\psi_+ = a\|xy>\alpha + b\|yz>\beta + ic\|xz>_\beta$) | $t_{2g}^6\,e_g$ ($\psi_+ = \|3z^2-r^2>\alpha$) |

$g_{xx} = 2(a^2 - b^2 + c^2 + 2kac)$

$g_{yy} = 2(a^2 + b^2 - c^2 + 2kab)$

$g_{zz} = 2(a^2 - b^2 - c^2 - 2kbc)$

$A_{xx} = 2g_n\beta_n\beta<r^{-3}>N^2[4ac - (\kappa+1/7)(a^2-b^2+c^2) + 3(1-2b(a+c))/7]$

$A_{yy} = 2g_n\beta_n\beta<r^{-3}>N^2[4ab - (\kappa+1/7)(a^2+b^2-c^2) + 3(1-2c(a+b))/7]$

$A_{zz} = -2g_n\beta_n\beta<r^{-3}>N^2[4bc - (\kappa+1/7)(-a^2+b^2+c^2) + 3(1-2a(b+c))/7]$

$q_{xx} = \dfrac{e^2Q}{6}<r^{-3}>N^2(1-R)[2a^2-4b^2+2c^2]/7$

$q_{yy} = \dfrac{e^2Q}{6}<r^{-3}>N^2(1-R)[2a^2+2b^2-4c^2]/7$

$q_{zz} = \dfrac{e^2Q}{6}<r^{-3}>N^2(1-R)[-4a^2+2b^2+2c^2]/7$

$g_\perp = 2.0$

$g_\| = 2.0$

$A_\perp = 2g_n\beta_n\beta<r^{-3}>N^2[-2/7-\kappa]$

$A_\| = 2g_n\beta_n\beta<r^{-3}>N^2[+4/7-\kappa]$

$q_\perp = \dfrac{e^2Q}{6}<r^{-3}>N^2(1-R)(2/7)$

$q_\| = \dfrac{e^2Q}{6}<r^{-3}>N^2(1-R)(-4/7)$

APPENDIX B - Notation Transformation

$$\psi_+ = a|xy>\alpha + b|yz>\beta + ic|xz>\beta$$

$$\psi-' = A|1\alpha> + B|\zeta\beta> + C|-1\alpha>$$

(see ref. 4)

$$a = B$$

$$b = (A-C)/\sqrt{2}$$

$$c = (A+C)/\sqrt{2}$$

APPENDIX C - Rhombic Field Parameters from a, b, c.

$$B/\lambda = \frac{1}{6}\left[\frac{a+b}{c} + \frac{b+c}{a} - \frac{2(a+c)}{b}\right]$$

$$A/\lambda = B/\lambda - \frac{1}{2}\left[\frac{b+c}{a} - \frac{a+c}{b}\right]$$

$$C = -A-B$$

RELATIVISTIC EFFECTS IN HYPERFINE INTERACTIONS[*]

B. D. Dunlap

Argonne National Laboratory
Argonne, Illinois 60439

The treatment of the theory of hyperfine interactions beginning initially from a relativistic viewpoint yields a number of results which differ from those of the non-relativistic approach. For both dipole and quadrupole interactions, one finds the necessity for different $\langle r^{-3} \rangle$ values for each in the terms of the Hamiltonian. In the magnetic dipole case, an additional term is present which has the appearance of a core polarization contribution to the hyperfine field, although it arises strictly from the open shell electrons. In the electric quadrupole case, two additional interactions appear which have no non-relativistic analogue. The relativistic approach will be briefly summarized and applications to the lanthanide and actinide series reviewed. Of particular interest will be the consideration of experimental values of $\langle r^{-3} \rangle$ and of core polarization fields.

INTRODUCTION

In discussions of relativity among those who use the Mössbauer effect to measure hyperfine interactions, attention has been placed almost entirely on the rather dramatic effects that occur with the isomer shift in heavy elements. This particular problem has been reviewed extensively

[*]Work performed under the auspices of the USAEC.

recently by Kalvius, and both the important phenomena as
well as the frequent pitfalls are discussed there.[1] Less
common is any mention of the effects of relativity on dipole
and quadrupole hyperfine interactions, except for an occa-
sional reference to the fact that values of $<r^{-3}>$ may be
altered when relativistic effects are properly accounted
for. Actually, a complete theoretical formulation[2-4] of
the hyperfine interaction problem which begins initially
from a relativistic viewpoint has been available for several
years, however the theory has been applied in detail to
only a few cases. A principal reason for this is that
sensible calculations require knowing radial integrals that
must be obtained from difficult relativistic self-consistent-
field computer calculations. Because of their complexity,
these programs have been largely used by only a select few,
however a number of results are now becoming available. In
particular, relativistic radial integrals have been recently
reported for a very large number of lanthanide[5] and
actinide[6] ions, making it possible to treat these two
series uniformly in a physically more correct fashion.

        The aim of this paper is to gather together some of
these results, both theoretical and experimental, and bring
them to the attention of a wider audience. We will first
briefly indicate the types of problems involved in calculating hyperfine interactions from solutions of the Dirac
equation as opposed to solutions of the Schrödinger equation,
and how these difficulties have been met. Using those
theoretical results and the available radial integrals, we
will then consider some cases in the actinide and lanthanide
series to see specifically the relativistic effects, how
important they may be, and how they relate to various
experiments.

          THE DIRAC EQUATION AND HYPERFINE INTERACTIONS

        In order to get an initial orientation and to under-
stand some of the problems we will encounter, we should
recall a few basic facts concerning the Dirac equation and
its solutions. If we begin from special relativity energy
relations and try to infer a relativistic wave equation,
then the natural choice for a free electron is

$$[\underset{\sim}{p}^{2}c^{2} + m^{2}c^{4}] \ \psi = E^{2}\psi \qquad\qquad (1)$$

This presents several difficulties, however, associated with the presence of E^2 which causes the wave equation to be of second order in time. Dirac[7] chose to linearize this by postulating the form

$$[\underset{\sim}{\alpha} \cdot \underset{\sim}{p} c + \beta m c^2] \psi = E\psi \qquad (2)$$

In order for Eqs. (1) and (2) to be compatible, α and β must be at least 4x4 matrices, and one can show that $\tilde{\alpha}$ is closely related to the spin angular momentum. Therefore spin is included in the problem from the beginning, whereas in non-relativistic theory it is tacked on after the orbital problem has been solved.

In a central field potential $V_c(r)$, the Dirac equation becomes

$$[\underset{\sim}{\alpha} \cdot \underset{\sim}{p} c + \beta m c^2 - eV_c(r)] \psi = E\psi \qquad (3)$$

This can be solved explicitly, however the presence of the 4x4 matrices causes the wave functions to have four components. These have the form

$$\psi_{jm} = \begin{pmatrix} \chi_{\ell jm} \ P(r)/r \\ \chi_{\ell' jm} \ iQ(r)/r \end{pmatrix} \qquad (4)$$

where $\ell' = \ell \pm 1$ for $j = \ell \pm 1/2$. The angular functions are given by

$$\chi_{\ell jm} = \sum_{\sigma} <1/2\ell jm|1/2\sigma\ell m - \sigma> Y_\ell^m \ \phi_{1/2}^\sigma \qquad (5)$$

where Y_ℓ^m is a spherical harmonic, $\phi_{1/2}^\sigma$ is a two component spin eigenfunction and $<|>$ denotes a Clebsch-Gordan coefficient. The two radial functions $P(r)$ and $Q(r)$ satisfy coupled differential equations and their explicit functional form depends, of course, on $V_c(r)$. For the many electron case, these should be solved by a Hartree-Fock type approach, and this problem has been discussed for Dirac electrons by Grant.[8] In the non-relativistic limit, the two radial equations can be combined to give the usual Schrödinger radial equation for a central field. In this limit one finds $Q(r) \ll P(r)$, so the upper and lower components of Eq. (4) are usually referred to as the large and small components respectively.

In order to find hyperfine interaction matrix elements, one uses the above wave-functions in a first order perturbation theory calculation, just as in non-relativistic

theory. However, there are several important features of
the relativistic wave-functions that make this calculation
more complicated:

(i)   Non-relativistic wave-functions have a radial
function, $R(r)$, which is used in the calculation of radial
integrals such as

$$<r^{-3}> = \int_{0}^{\infty} |R(r)|^2 \, r^{-3} dr \qquad (6)$$

Relativistic wave-functions, however, have the two radial
parts, $P(r)$ and $Q(r)$. The way in which these should be
combined to form an expression analogous to Eq. (6) is not
completely obvious.

(ii)   In non relativistic theory, spin is added in
an ad-hoc way to solutions of the Schrödinger equation.
The radial function, $R(r)$, is therefore independent of the
spin. In the Dirac equation, where spin enters as an
intrinsic part of the whole problem, the radial parts of
Eq. (4) will be different for electrons having $j = \ell + 1/2$
or $j = \ell - 1/2$. This adds another complication to integrals
such as Eq. (6), since the matrix elements $<r^{-3}>_{++} \equiv$
$<\ell + 1/2|r^{-3}|\ell + 1/2>$, $<r^{-3}>_{--} \equiv <\ell - 1/2|r^{-3}|\ell - 1/2>$ and
$<r^{-3}>_{+-} \equiv <\ell + 1/2|r^{-3}|\ell - 1/2>$ will all be different.

(iii)   Because of the presence of both ℓ and ℓ' in
Eq. (4), the relativistic wave-functions are not eigen-
functions of the orbital angular momentum $\underset{\sim}{I}$. Thus, in a
many electron problem these wave functions are most easily
combined to form states having jj coupling. However, for
most cases of practical interest, the electrons are much
nearer to LS coupling.

Because of these various problems, it is not possible
to simply extend the non-relativistic theory of hyperfine
interactions into the relativistic regime by an intuitive
process. It is necessary to do the problem over from the
beginning. This has been done in detail for a single
Dirac electron by Schwartz,[2] using a standard multipole
expansion, so the appropriate one-electron matrix elements
are known. In particular, he solves problem (i) above by
deriving the appropriate form of the radial integrals for
all multipolarities of the interactions. Schwartz finds,
in fact, that the required integrals are different for the

magnetic and electric cases, so a single $<r^{-3}>$ cannot be applied to both.

The other difficulties have been remedied by Sandars and Beck.[3] First, they consider the formation of a relativistic many-electron LS coupled state out of single-electron jj coupled wave-functions. All matrix elements are thus reduced to a sum of matrix elements between single particle states like Eq. (4). They then show how to construct an effective hyperfine Hamiltonian which has matrix elements between non-relativistic LS coupled states that are equal to those of the actual hyperfine Hamiltonian between LS coupled relativistic states. In this process they show the proper way to combine the integrals $<r^{-3}>_{++}$, $<r^{-3}>_{+-}$ and $<r^{-3}>_{--}$ for the various possible interactions that occur. Once this has been done, one no longer has the complexity of making calculations directly from Eq. (4), except that one must still obtain a few one-electron radial integrals.

The method of Sandars and Beck can be carried through in a straightforward and elegant way by the use of tensor operator techniques.[9] However, it is not possible in the present space to both present the details of that computation and to also discuss its experimental implications. We therefore choose to give some results from that paper and then apply these results to the lanthanide and actinide series, where relativistic effects may be noticeable. Thus the reader will be asked on occasion to accept some obscure expressions which we will try to make plausible, but is referred to the original papers for details.

MAGNETIC DIPOLE INTERACTIONS

As in classical mechanics, we may account for the presence of a magnetic field having a vector potential A by making the replacement $p \rightarrow p + (e/c) A$. This introduces into Eq. (3) a perturbation $\mathcal{H}' = (e/c) \alpha \cdot A$. Using the one-electron matrix elements of Schwartz[2] and calculating the many electron problem in LS coupling, Sandars and Beck find[3] that the hyperfine interaction for the dipole case may be written

$$\mathcal{H}_{MD} = g_n \mu_n I \cdot H_{hf} \tag{7}$$

where $g_n \mu_n I$ is the nuclear magnetic moment and the magnetic hyperfine field, H_{hf}, is given by

$$H_{hf} = (\mu_0 \mu_B / 2\pi) \sum_i \{ \underset{\sim}{\ell}_i \langle r_\ell^{-3} \rangle$$

$$+ [3\underset{\sim}{r}_o (\underset{\sim}{r}_o \cdot \underset{\sim}{s}_i) - \underset{\sim}{s}_i] \langle r_{sd}^{-3} \rangle + \underset{\sim}{s}_i \langle r_s^{-3} \rangle \}. \tag{8}$$

Here, $\underset{\sim}{\ell}_i$ and $\underset{\sim}{s}_i$ are the orbital and spin angular momenta of the ith electron, and $\underset{\sim}{r}_o$ is a unit vector connecting the nucleus and the ith electron. The first two terms of Eq. (8) correspond to the usual non-relativistic result, giving the hyperfine field from orbital and spin-dipolar interactions. In addition, we see two ways in which relativity alters the hyperfine field:

(1) There is a third term which is proportional to $\underset{\sim}{s}_i$ and so has the appearance of a contact interaction or a core polarization. However, it is purely relativistic in origin. True polarization of the core has not been treated in this approach, and so does not appear in Eq. (8).

(2) There are different radial parameters for each of the three different interactions. This occurs because the one electron basis states with $j = \ell + 1/2$ and $j = \ell - 1/2$ have different radial distributions, and these components are weighted differently by the different interactions.

Specifically, the radial integrals are[10]

$$\langle r_\ell^{-3} \rangle = [2\ell(\ell + 1)F_{++} + 2\ell(\ell + 1)F_{--} + F_{+-}]/(2\ell + 1)^2 \tag{9a}$$

$$\langle r_{sd}^{-3} \rangle = [-4\ell(\ell + 1)(2\ell - 1)F_{++} + 4\ell(\ell + 1)(2\ell + 3)F_{--}$$
$$- (2\ell + 3)(2\ell - 1)F_{+-}]/3(2\ell + 1)^2 \tag{9b}$$

$$\langle r_s^{-3} \rangle = 4[\ell(\ell + 1)^2 F_{++} - \ell^2(\ell + 1)F_{--}$$
$$- \ell(\ell + 1)F_{+-}]/3(2\ell + 1)^2 \tag{9c}$$

As before, ℓ is the single electron orbital momentum
(e.g., $\ell = 3$ for f electrons). The quantities $F_{jj'}$ are
one-electron radial integrals given by

$$F_{jj'} = - [2\hbar/mc(K + K' + 2)] \cdot$$

$$\int_0^\infty [P_j(r)Q_{j'}(r) + P_{j'}(r)Q_j(r)]r^{-2}dr$$

(10)

with $K = \pm (j + 1/2)$ for $j = \ell \pm 1/2$. In Eqs. (9), the
subscripts $\pm$ denote $j,j' = \ell \pm 1/2$. In the non-relativistic
limit one can show that $F_{++} = F_{+-} = F_{--} = <r^{-3}>$. Thus from
Eqs. (9) we see that $<r_\ell^{-3}> = <r_{sd}^{-3}> = <r^{-3}>$ and $<r_s^{-3}> = 0$.
Eq. (8) then becomes the common expression for the hyperfine
field.*

    The $F_{jj'}$ have recently been calculated by W. Burton
Lewis and co-workers for a large number of lanthanide[5]
and actinide[6] ions using a Dirac-Slater (i.e., relativistic
self-consistent-field with Slater exchange) computer program.
Using those numbers, Eqs. (9) may be evaluated. The results
are given in Table I for all the trivalent lanthanides plus
Eu^{2+}, and in Table II for a large number of ions in the
first half of the actinide series.

    For a state obeying Hund's rules, Eq. (10) can be
expressed in a relatively simple form. If we specify the
state at magnetic saturation by the quantum numbers L,S,J
and $J_z = J$, then one finds[11]

$$H_{hf} = a [(2 - g_J) <r_\ell^{-3}> + \alpha_{sd} <r_{sd}^{-3}> + (g_J - 1) <r_s^{-3}>]J$$

$$= H_\ell + H_{sd} + H_s$$

(11)

with
$$g_J = 1 + [S(S + 1) - L(L + 1) + J(J + 1)]/2J(J + 1) \quad (12)$$
and

$$\alpha_{sd} = \frac{(2\ell+1-4S)2}{(2\ell-1)(2\ell+3)(2L-1)S} \{2L(L + 1)(g_J - 1)$$
$$-3(2 - g_J)[J(J + 1) - L(L + 1) - S(S + 1)]\}$$

(13)

---

*In the literature, the spin-dipolar radial integral here
denoted $<r_{sd}^{-3}>$ is frequently called $<r_{sc}^{-3}>$. This notation
arises from tensor operator formalism.

Table I. Dipole Radial Parameters and Hyperfine Field
          Contributions in Lanthanide Ions

	$\langle r_\ell^{-3}\rangle$	$\langle r_{sd}^{-3}\rangle$	$\langle r_s^{-3}\rangle$	H_ℓ	H_{sd}	H_s	H_{hf}
	(a.u.)	(a.u.)	(a.u.)	(kOe)	(kOe)	(kOe)	(kOe)
$Ce^{3+}4f^1$	4.338	4.566	-0.108	1551	326	5	1882
$Pr^{3+}4f^2$	4.909	5.179	-0.127	2948	299	13	3260
$Nd^{3+}4f^3$	5.498	5.815	-0.149	3939	139	23	4101
$Pm^{3+}4f^4$	6.109	6.481	-0.176	4280	-138	35	4177
$Sm^{3+}4f^5$	6.743	7.174	-0.203	3615	-444	45	3216
$Eu^{2+}4f^7$	6.776	7.320	-0.259	0	0	-113	-113
$Eu^{3+}4f^6$	7.403	7.904	-0.236	0	0	0	0
$Gd^{3+}4f^7$	8.099	8.673	-0.270	0	0	-118	-118
$Tb^{3+}4f^8$	8.826	9.481	-0.308	3312	158	-116	3354
$Dy^{3+}4f^9$	9.585	10.333	-0.352	5995	431	-110	6316
$Ho^{3+}4f^{10}$	10.374	11.221	-0.398	7787	187	-100	7874
$Er^{3+}4f^{11}$	11.197	12.156	-0.451	8404	-217	- 85	8102
$Tm^{3+}4f^{12}$	12.054	13.134	-0.508	7540	-548	- 64	6928
$Yb^{3+}4f^{13}$	12.946	14.161	-0.572	4859	-591	- 36	4232

If the radial parameters are given in atomic units (a.u.),
that is, in multiples of a_o^{-3} where a_o is the Bohr radius,
then a has the value

$$a = 0.1251 \times 10^6 \text{ Oe/a.u.}$$

Again, Eq. (11) differs from the non-relativistic result[12]
in the different values of $\langle r^{-3}\rangle$ and in the presence of the
term containing $\langle r_s^{-3}\rangle$. Tables I and II give the results of
evaluating Eq. (11) for the various lanthanide and actinide
ions.

Let us now consider a few experiments and try to evaluate
the reliability of the relativistic $\langle r^{-3}\rangle$ values. It has

Table II. Dipole Radial Parameters and Hyperfine Field
Contributions in Actinide Ions

	$\langle r_\ell^{-3} \rangle$	$\langle r_{sd}^{-3} \rangle$	$\langle r_s^{-3} \rangle$	H_e	H_{sd}	H_s	H_{hf}^*
	(a.u.)	(a.u.)	(a.u.)	(kOe)	(kOe)	(kOe)	(kOe)
$U^{2+}5f^4$	4.823	5.850	-0.481	3379	-124	96	3351
$U^{3+}5f^3$	5.513	6.469	-0.442	3950	154	68	4172
$U^{4+}5f^2$	6.164	7.105	-0.431	3972	356	43	4371
$U^{5+}5f^1$	6.798	7.739	-0.427	2430	485	19	2934
$Np^{3+}5f^4$	6.221	7.314	-0.506	4358	-155	101	4304
$Np^{4+}5f^3$	6.868	7.941	-0.490	4921	189	75	5185
$Np^{5+}5f^2$	7.500	8.574	-0.487	4503	496	49	5048
$Np^{6+}5f^1$	8.125	9.206	-0.486	2904	658	22	3584
$Pu^{3+}5f^5$	6.942	8.189	-0.574	3722	-430	128	3420
$Pu^{4+}5f^4$	7.588	8.805	-0.556	5316	-187	111	5240
$Pu^{5+}5f^3$	8.222	9.432	-0.549	5891	225	84	6200
$Pu^{6+}5f^2$	8.855	10.071	-0.547	5317	582	55	5954
$Am^{2+}5f^7$	7.035	8.527	-0.695	0	0	-304	-304
$Am^{3+}5f^6$	7.678	9.081	-0.647	0	0	0	0
$Am^{4+}5f^5$	8.327	9.701	-0.628	4309	-601	140	3848
$Am^{5+}5f^4$	8.964	10.315	-0.613	6280	-219	123	6184
$Am^{6+}5f^3$	9.604	10.966	-0.613	6881	262	94	7237

*In actinide materials, significant deviations from the
Hund's rule state may occur due to large crystal field
interactions. See Ref. (19).

been previously noted that the original Hartree-Fock values $\langle r^{-3}\rangle_{HF}$ calculated by Freeman and Watson[13] for the tri-valent lanthanides are significantly larger than empirical values obtained by Bleaney.[14] Judd has shown[15] that configuration interaction can be invoked to account for the difference. Alternatively, Watson and Freeman suggested[16] that the differences may arise simply because the original calculation was non-relativistic. It would therefore be of interest to compare the values in Table I with those of Bleaney, however this cannot be done directly because of the three different values of $\langle r^{-3}\rangle$ given by the relativistic calculation. To circumvent this problem, we will define an effective value $\langle r^{-3}\rangle_{eff}$ by using a non-relativistic expression for the hyperfine field:

$$H_{hf} = a[(2 - g_J)J + \alpha_{sd}J] \langle r^{-3}\rangle_{eff} \qquad (14)$$

If we now find the values of $\langle r^{-3}\rangle_{eff}$ required to give the H_{hf} listed in Table I for the various lanthanides, then all the relativistic effects will be absorbed into the one radial parameter. The results of such a calculation are shown in Fig. 1, where they are compared both with the experimental values and with the original Hartree-Fock results. One sees that essentially all the discrepancy has been resolved by the inclusion of relativity. These calculated values may be changed somewhat if one includes configuration interaction, core polarization and departure from the pure Hund's rule state, however these effects would appear to be considerably less important.

While this is a satisfying result, it nonetheless does not provide the best test for the calculated $\langle r^{-3}\rangle$ values since the subtleties of the relativistic effects are lost. In one case, however, a detailed comparison with experiment is possible. Atomic beam experiments have been performed and hyperfine interactions obtained[17] for several different J manifolds in Sm. This provides enough data to determine the three radial parameters of Eq. (8) experimentally. The values obtained for $\langle r_l^{-3}\rangle$, $\langle r_{sd}^{-3}\rangle$ and $\langle r_s^{-3}\rangle$ were 6.390 a.u., 6.513 a.u. and -0.208 a.u. The Dirac-Slater calculation of Lewis et. al.[5-6] for atomic Sm gave 6.049 a.u., 6.554 a.u. and -0.240 a.u. It is especially satisfying that agreement is obtained both for the sign and the magnitude of the purely relativistic quantity $\langle r_s^{-3}\rangle$. A similar analysis has been given by Rosén.[18]

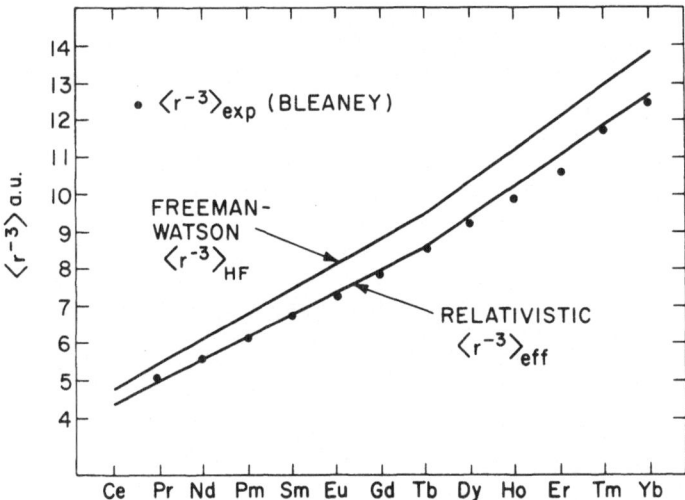

Fig. 1. Experimental and calculated $\langle r^{-3} \rangle$ values for tri-valent lanthanides. The relativistic $\langle r^{-3} \rangle_{eff}$ is calculated for the pure Hund's rule state as described in text.

Using the Dirac-Slater values, Lewis[5] has also calculated hyperfine interaction constants for a number of rare-earths using known intermediate coupling wavefunctions and compared these with various atomic-beam experiments. Agreement is generally obtained to about 1%. In view of these combined results, one may feel justified in relying on the validity of the calculated relativistic $\langle r^{-3} \rangle$ values for the lanthanides.

In the actinide series, one expects relativistic effects to be considerably more important because the atoms are substantially heavier, however much less experimental data is available. In addition, crystal field effects are quite important in this series, often causing large deviations from the Hund's rule state and so greatly complicating the data analysis.[19] Recently, EPR data has been obtained[20] for Pu^{3+} in CaF_2, SrF_2 and BaF_2. The data was analyzed including a mixing of the first excited J manifold ($J = 7/2$) into the lowest J manifold ($J = 5/2$). With the assumptions $\langle r_\ell^{-3} \rangle = \langle r_{sd}^{-3} \rangle = \langle r^{-3} \rangle_{exp}$ and $\langle r_s^{-3} \rangle = 0$, a value of $\langle r^{-3} \rangle_{exp} = 7.57 \pm 0.57$ a.u. was obtained. If we use the calculated hyperfine energy matrix elements from that work and the radial parameters of Table II, and parameterize in a way similar to Eq. (16), then a calculated value of $\langle r^{-3}\rangle_{eff}$ = 7.04 a.u. is found, in adequate agreement with experiment. Lewis et al.[6] have reanalyzed existing atomic beam data for ^{239}Pu and find substantial agreement with the calculated $\langle r^{-3}\rangle$ values, although the experimental errors are larger due to the low nuclear moment. A number of related experiments are discussed in Ref. (6).

While the experimental results are not so abundant for the actinides as for the lanthanides, it nonetheless seems clear that the relativistic calculations are again producing sensible results. It must be strongly emphasized that non-relativistic self-consistent-field calculations produce values in clear disagreement with these experimental results. For example, a Hartree-Fock-Slater calculation[21] for the Pu^{3+} case discussed above gives $\langle r^{-3}\rangle$ = 9.00 a.u. In general, the non-relativistic $\langle r^{-3}\rangle$ values[21] are much larger than the relativistic ones. Thus, for the actinides it is quite important to include the relativistic effect.

We turn now to the second relativistic effect, namely the term of Eq. (8) which is proportional to s_i. For a state in which J is a good quantum number then this will be, as in Eq. (13),

$$H_s = 125(g_J - 1) \langle r_s^{-3}\rangle \text{ kOe.} \qquad (15)$$

This has exactly the form generally assumed for core polarization. For example, in the lanthanides one has frequently taken

$$H_{core} = -90(g_J - 1)J \text{ kOe} \qquad (16)$$

We must emphasize, however, that H_s arises strictly from the open shell electrons, and occurs because electrons with spin projection + 1/2 and - 1/2 in the open shell have different radial distributions. Actual core polarization has a rather different physical origin, but, because of the similarity of Eqs. (17) and (18), the two contributions are experimentally indistinguishable. In the following we will denote the experimental "core polarization" hyperfine fields by

$$H_{core}^{exp} = H_s + H_{core} \qquad (17)$$

For the trivalent lanthanides, $<r_s^{-3}>$ varies from -0.1 to
-0.6 a.u. From Eqs. (17) and (18) we see that H_s can be
comparable in magnitude to H_{core} and has the same sign.
Therefore any discussion of measured values H_{core}^{exp} must take
the relativistic term into consideration. This will be of
special importance in ions with a half-filled shell such
as Eu^{2+} or Am^{2+}, where the ground state is $^8S_{7/2}$. In these
cases both the orbital and spin-dipolar contributions
vanish, and the entire hyperfine field is due to the terms
of Eq. (19).

A few values of H_{core}^{exp} are available, both in free
atoms from atomic beam work, and in ions from EPR and
Mössbauer effect measurements. These have been summarized
in Table III and, for comparison, the corresponding values
of H_s from Tables I and II are included. A note of warning
should be offered concerning this comparison: one should
properly correct for departures from pure LS coupling, and
we have not done so for the H_s values. However the changes
due to such effects are rather small for the cases given
here. The Table shows quite a remarkable result. For the
un-ionized atoms, where the 7s shell is filled, the observed
values can be entirely attributed to H_s, the relativistic
term. Thus in the atoms one infers that true core polarization is small. In the ionized atoms, where the 7s shell
is empty, H_s is still a significant part of the measured
fields, however it is clear that here actual core polarization does take place. Easley[26] has suggested that the
small H_{core} in the atoms is due to two things: (1) when
the 7s shell is filled, it is not possible to admix 7s
orbitals into the core orbitals, so a principal excitation
contributing to the core polarization is removed. (2) When
the 7s shell is filled, it too will be polarized, and this
can make a contribution to the hyperfine field which is of
opposite sign to that of the inner shells.

The importance of H_s has some significance for previous
calculations of H_{core}. For Eu^{2+}, Eq. (16) gives $H_{core} \simeq$
-300 kOe. While one has never relied on this value for
great accuracy, it was nonetheless a satisfying result since
measured hyperfine fields in Eu^{2+} generally fall in the
range of -300 to -350 kOe. However, we see from Table I
that about -100 kOe of this is due to H_s, and therefore
Eq. (16) must over-estimate the actual value of H_{core}. This
is in line with a more recent Hartree-Fock calculation[28]

Table III. Effective "Core Polarization" Fields in
            Lanthanide and Actinides

	$H_s/(g_J - 1)J$ (kOe)	$H_{core}^{exp}/(g_J - 1)J$ (kOe)	Ref.
Eu $4f^7 6s^2$	-30	-29	22
Am $5f^7 7s^2$	-97	-70	23
Pu $5f^6 7s^2$	-89	-98 ± 58	24
$Eu^{2+} 4f^7$	-32	-95 ± 10	25
$Am^{2+} 5f^7$	-80	-550 ± 55	26
		-630	27
$Pu^{3+} 5f^5$	-72	-730 ± 60	20

which gave an almost vanishing core-polarization for Yb^{3+},
in disagreement with the simple approach expressed by Eq. (16).
We therefore draw two conclusions: (i) Eq. (16) is not
really valid for the lanthanide series, and (ii) detailed
discussions of the hyperfine fields of the half-filled shell
ions must not ignore H_s.

While on the subject of core-polarization, one should
note that the observed values for the actinides ions give

$$H_{core}^{exp} = - (630 \pm 100)(g_J - 1)J \text{ kOe} \qquad (18)$$

over the admittedly limited number of configurations available.
This is an order of magnitude larger than is seen in the
lanthanides and, in fact, makes a major contribution to the
hyperfine field in most cases. Thus, in contrast to the
lanthanides, one cannot ignore core-polarization for actinide
ions even in first order.[29] From Table II we see that the
expression $<r_s^{-3}> = (0.55 \pm 0.07)$a.u. covers all the listed
ions. This gives $H_s \simeq - 70(g_J - 1)J$ kOe. Therefore one can
include core polarization in an approximate way by adding

$$H_{core} = -(560 \pm 100)(g_J - 1)J \text{ kOe} \qquad (19)$$

to the values of H_{hf} listed in Table II. However, the wide-spread validity of Eq. (19) has yet to be demonstrated.

ELECTRIC QUADRUPOLE INTERACTIONS

We turn now to quadrupole interactions. Unfortunately in this case the results have a somewhat more foreboding appearance. If the nuclear state is specified by the quantum numbers I and I_z, then the hyperfine interaction is

$$\mathcal{H}_{EQ} = \frac{e^2 qQ}{4I(2I-1)} [3I_z^2 - I(I+1)] \qquad (20)$$

Where Q is the nuclear quadrupole moment. If the electron is in an LS coupled state specified as $|SLJ \, J_z\rangle$ then one knows[30] from the Wigner-Eckart theorem that the electric field gradient arising from valence electrons can be written

$$eq = (-1)^{J-J_z} \begin{pmatrix} J & 2 & J \\ -J_z & 0 & J_z \end{pmatrix} \langle J\| \, T_e^{(2)} \, \|J\rangle. \qquad (21)$$

Here, ( ) denotes a 3 - j symbol, $T_e^{(2)}$ is some appropriate combination of electron operators, and $\langle\| \, \|\rangle$ denotes a reduced matrix element which should be calculated by the rules of tensor operators.[9] Explicit calculation of $T_e^{(2)}$ has been carried out by Schwartz[2] for the single electron problem and extended to the LS coupled many-electron case by Sandars and Beck.[10] The result is given in very general terms by the following

$$T_e^{(2)} = 2\sqrt{\frac{\ell(\ell+1)(2\ell+1)}{(2\ell+3)(2\ell-1)}} \sum_i \left[\sqrt{2} \, U_i^{(02)2} \langle r^{-3}\rangle_{02} \right.$$

$$\left. -\frac{3}{5}\sqrt{14(\ell+2)(\ell-1)} \, U_i^{(13)2} \langle r^{-3}\rangle_{13} - \frac{6}{5}\sqrt{(2\ell+3)(2\ell-1)} \, U_i^{(11)2} \langle r^{-3}\rangle_{11} \right] \qquad (22)$$

The quantities $U_i^{(k_s k_\ell)k}$ are unit tensor operators of order k_s in spin space, k_ℓ in orbital space and k in the combined spin-orbital space. When (24) has been substituted into (23), then we obtain an expression for the field gradient which we can denote symbolically by

$$q = q_{02} + q_{13} + q_{11} \qquad (23)$$

The radial matrix elements are given by Sandars and Beck to be

$$\langle r^{-3}\rangle_{02} = [(\ell+2)(2\ell-1)R_{++} + (\ell-1)(2\ell+3)R_{--} + 6R_{+-}]/(2\ell+1)^2$$
(24a)

$$\langle r^{-3}\rangle_{11} = [-(\ell+2)R_{++} + (\ell-1)R_{--} + 3R_{+-}]/(2\ell+1)^2 \qquad (24b)$$

$$\langle r^{-3}\rangle_{13} = [(2\ell-1)R_{++} - (2\ell+3)R_{--} + 4R_{+-}]/(2\ell+1)^2 \qquad (24c)$$

where

$$R_{jj'} = \int_0^\infty [P_j(r)P_{j'}(r) + Q_j(r)Q_{j'}(r)]r^{-3}dr \qquad (25)$$

and a notation similar to Eq. (10) has been used. In the non-relativistic limit one sees that $\langle r^{-3}\rangle_{02} = \langle r^{-3}\rangle$ and $\langle r^{-3}\rangle_{13} = \langle r^{-3}\rangle_{11} = 0$. Values for these radial parameters are given in Tables IV and V for the lanthanide and actinide ions previously discussed, as obtained using the Dirac-Slater matrix elements of Lewis et al.[5-6]

The detailed evaluation of Eq. (21) is quite cumbersome in general. However, the operator $U^{(02)2}$ can be shown to be proportional to the spherical harmonic $Y_{20} \sim 3\cos^2\theta - 1$. Thus this term corresponds to the usual non-relativistic field gradient, and can be written

$$q_{02} = \langle J\|\alpha\|J\rangle [3J_z^2 - J(J + 1)] \langle r^{-3}\rangle_{02} \qquad (26)$$

The values of $\langle J\|\alpha\|J\rangle$ have been calculated by Stevens for the f^n Hund's rule states,[31] and are included in Tables IV and V. The other two terms in eq are purely relativistic in origin, arising from distortions of the electron distribution due to the different radial wave functions of the electrons having $j = \ell + 1/2$ and $j = \ell - 1/2$. The operators $U^{(13)2}$ and $U^{(11)2}$ have no analogues in non-relativistic hyperfine theory. However, from the small values of $\langle r^{-3}\rangle_{13}$ and $\langle r^{-3}\rangle_{11}$ in Tables IV and V, we may expect that their contribution will not be large.

We may get some idea of the size of the various terms by considering a simple case, namely an f^1 configuration in the state $|S = 1/2, L = 3, J = 5/2, J_z = 5/2\rangle$. This would

Table IV. Quadrupole Radial Parameters and Values of
$\langle J\|\alpha\|J\rangle$ for Lanthanide Ions

| | $\langle r^{-3}\rangle_{02}$ | $\langle r^{-3}\rangle_{13}$ | $\langle r^{-3}\rangle_{11}$ | $\langle J\|\alpha\|J\rangle$ |
	(a.u.)	(a.u.)	(a.u.)	
$Ce^{3+}4f^1$	4.348	-0.017	0.008	-0.0571
$Pr^{3+}4f^2$	4.821	-0.020	0.010	-0.0210
$Nd^{3+}4f^3$	5.512	-0.024	0.011	-0.0064
$Pm^{3+}4f^{11}$	6.126	-0.028	0.013	0.0077
$Sm^{3+}4f^5$	6.762	-0.032	0.015	0.0413
$Eu^{2+}4f^7$	6.795	-0.040	0.019	0
$Eu^{3+}4f^6$	7.426	-0.038	0.018	0
$Gd^{3+}4f^7$	8.127	-0.044	0.020	0
$Tb^{3+}4f^8$	8.857	-0.050	0.023	-0.0101
$Dy^{3+}4f^9$	9.619	-0.057	0.027	-0.0063
$Ho^{3+}4f^{10}$	10.414	-0.065	0.030	-0.0022
$Er^{3+}4f^{11}$	11.243	-0.073	0.034	0.0025
$Tm^{3+}4f^{12}$	12.105	-0.082	0.039	0.0101
$Yb^{3+}4f^{13}$	13.003	-0.129	0.043	0.0317

be, for example, the Hund's rule case for Ce^{3+} or Np^{6+}.
For this single electron problem we can evaluate Eq. (21)
in closed form to obtain

$$eq = b[4/7\ \langle r^{-3}\rangle_{02} - 12/7\ \langle r^{-3}\rangle_{13} + 8/7\ \langle r^{-3}\rangle_{11}] \quad (27)$$

with $b = 9.75 \times 10^{17}$ V/cm^{2}a.u. Using the values in Tables
IV and V, this gives

$$eq = \begin{cases} 2.422 + .028 + .009 \text{ V/cm}^2 \text{ for } Ce^{3+} \\ 4.595 + .159 + .047 \text{ V/cm}^2 \text{ for } Np^{6+} \end{cases} \quad (28)$$

Table V.   Quadrupole Radial Parameters and Values of $\langle J \| \alpha \| J \rangle$ for Actinide Ions

	$\langle r^{-3} \rangle_{02}$ (a.u.)	$\langle r^{-3} \rangle_{13}$ (a.u.)	$\langle r^{-3} \rangle_{11}$ (a.u.)	$\langle J \| \alpha \| J \rangle^{*}$
$U^{2+}5f^{4}$	4.879	-0.081	0.037	0.0077
$U^{3+}5f^{3}$	5.584	-0.078	0.036	-0.0064
$U^{4+}5f^{2}$	6.249	-0.079	0.036	-0.0210
$U^{5+}5f^{1}$	6.895	-0.081	0.036	-0.0571
$Np^{3+}5f^{4}$	6.305	-0.090	0.041	0.0077
$Np^{4+}5f^{3}$	6.966	-0.091	0.041	-0.0064
$Np^{5+}5f^{2}$	7.611	-0.093	0.041	-0.0210
$Np^{6+}5f^{1}$	8.248	-0.095	0.042	-0.0571
$Pu^{3+}5f^{5}$	7.041	-0.103	0.046	0.0413
$Pu^{4+}5f^{4}$	7.700	-0.103	0.046	0.0077
$Pu^{5+}5f^{3}$	8.347	-0.105	0.047	-0.0064
$Pu^{6+}5f^{2}$	8.994	-0.108	0.048	-0.0210
$Am^{2+}5f^{7}$	7.132	-0.120	0.054	0
$Am^{3+}5f^{6}$	7.791	-0.117	0.053	0
$Am^{4+}5f^{5}$	8.454	-0.117	0.052	0.0413
$Am^{5+}5f^{4}$	9.108	-0.118	0.053	0.0077
$Am^{6+}5f^{3}$	9.760	-0.120	0.054	-0.0064

*In actinide materials, significant deviations from the Hund's rule state may occur due to large crystal field interactions. See Ref. (19).

We see that the extra terms cause a change of $\sim 1\%$ for the lanthanide and $\sim 5\%$ for the actinide. In view of the uncertainties commonly associated with the values of quadrupole moments, the uncertainty of knowing Sternheimer factors, the difficulty of assessing the lattice field gradient, these effects will usually be within experimental errors. Thus the major effect of relativity in quadrupole interactions is simply in giving the proper value for $\langle r^{-3}\rangle_{02}$.

In some cases, however, the effects of the relativistic corrections have been seen. As with the dipole interactions, they are most pronounced in the half filled shell, as in atomic Eu or Eu^{2+}. If the pure Hund's rule state $^8S_{7/2}$ is obtained for these cases, then one can show that all the quadrupole terms in Eq. (25) vanish. Nonetheless, in atomic Eu one observes[32] a small interaction $e^2qQ = -0.7012$ MHz. This is attributed to a breakdown of pure LS coupling which arises because the spin orbit-coupling admixes a bit of the state $^6P_{7/2}$ into the ground state. The amount of this admixture can be estimated from the known spin-orbit coupling constant and the energy separation between the S and P states. Now the L = 1 state will give rise to a quadrupole hyperfine interaction, however if one calculates the "non-relativistic" part $e^2q_{02}Q$ its sign is found to be positive (+ 1.87 MHz). In this case, then, it is necessary to include the relativistic correction to obtain agreement with experiment even with regard to sign. Estimates made by Evans et. al.[32] show that this does in principle solve the problem, giving a total interaction $e^2qQ = -0.57$ MHz. Unfortunately a later calculation by Coulthard,[33] with more accurate radial integrals, is in much worse agreement with experiment. Nonetheless, it does seem to be agreed that the presence of the relativistic correction terms has been demonstrated, although the details for this case remain unclear. A similar discussion has been given for atomic $Mn(^6S_{5/2})$ by Evans et al.[34] and for Eu in the configurations $4f^76s^2$, $4f^76s$ and $4f^7$ by Bordarier et al.[4]

CONCLUSION

We have seen that a treatment of hyperfine interactions from a relativistic viewpoint causes a number of modifications to the familiar non-relativistic expressions. Every term in the Hamiltonian is seen to require a different value

of $<r^{-3}>$. In addition, some extra terms appear. For magnetic dipole interactions there is one such new interaction
which has the appearance of a core polarization term,
although it arises solely from the open-shell electrons,
and which can play a significant role in understanding
measured core-polarization hyperfine fields. In electric
quadrupole interactions, two additional terms are present
which have no non-relativistic analogues.

Finally, there is one important point to be understood.
It is commonly assumed that relativistic effects will be
of importance only in the heaviest atoms. For the f electron systems discussed above, this is certainly true.
However, it is also true that the effects become more
pronounced earlier in the periodic table for lower angular
momentum states. For a systematic view of the trends involved, the reader is referred to the book of Kopfermann[35]
In particular, Casimir has shown[36] how to approximate
the radial integrals required, and values of the Casimir
correction factors are given in Ref. (35) for various
electrons as a function of Z.

For the present, we will only discuss a case where more
accurate information is available, namely, that of Sn.
Although this is only Z = 50, nonetheless the relativistic
effects are quite visible because the valence electrons are
in p states. Recently Childs[37] has performed atomic beam
measurements on five different levels in ^{117}Sn and ^{119}Sn.
This data was discussed both from relativistic and from
non-relativistic viewpoints, and in order to achieve self-
consistency in the large amount of data, it is essential
to include the relativistic effects. This necessity is
most easily seen by looking at the radial integrals. Using
the relativistic wave-function of Mann,[38] Childs cal-
culated the values $<r_\ell^{-3}> = 7.888$ a.u., $<r_{sd}^{-3}> = 10.126$ a.u.
and $<r_s^{-3}> = -0.880$ a.u. for neutral Sn. This, of course,
represents a significant departure from the non-relativistic
requirement of $<r_\ell^{-3}> = <r_{sd}^{-3}>$ and $<r_s^{-3}> = 0$. The $^3P_1$ state
is particularly interesting because the orbital and spin-
dipolar interactions would exactly cancel non-relativistically.
However, because of the large difference between $<r_\ell^{-3}>$ and
$<r_{sd}^{-3}>$, this is not the case. Also, about one-third of the
measured H_{core}^{exp} is found to arise from the rather large
$<r_s^{-3}>$. Thus it is very clear that in this case the relativistic effects cannot be considered merely as corrections.

ACKNOWLEDGMENTS

The author would like to express his gratitude to W. J. Childs who made a number of useful suggestions and who graciously provided the results of his Sn investigation prior to publication. W. B. Lewis and J. B. Mann of the Los Alamos Scientific Laboratory have contributed several useful discussions. Many of the results discussed here derive directly from their publications. W. B. Lewis has kindly provided the radial integrals for the lanthanides in advance of publication. I would also like to thank S. L. Ruby and G. J. Perlow for their reading of the manuscript.

REFERENCES

1. G. M. Kalvius, in "Proc. Int. Conf. Hyperfine Interactions Detected by Nuclear Radiation", Rehovot, Jerusalem, Israel, 1970, in press.

2. C. Schwartz, Phys. Rev. 97, 380 (1955).

3. P. G. H. Sandars and J. Beck, Proc. Roy. Soc. (London) A289, 97 (1965).

4. Y. Bordarier, B. R. Judd and M. Klapisch, Proc. Roy. Soc. (London) A289, 81 (1965).

5. W. Burton Lewis, in "Proc. XVIth Colloque Ampere: Magnetic Resonances and Related Phenomena", Bucharest, Romania, 1970, in press.

6. W. Burton Lewis, Joseph B. Mann, David A. Liberman and Don T. Cromer, J. Chem. Phys. 53, 809 (1970).

7. P. A. M. Dirac, "Quantum Mechanics" (Oxford Univ. Press, London, 1958).

8. I. P. Grant, Adv. Phys. 19, 747 (1970).

9. B. R. Judd, "Operator Techniques in Atomic Spectroscopy" (McGraw Hill, New York, 1963).

10. From Ref. (3), with the correction of a minor misprint.

11. J. C. Hubbs, R. Marrus, W. A. Nierenberg and J. L. Worcester, Phys. Rev. 109, 390 (1958).

12. For example, see J. Kondo, J. Phys. Soc. Japan 16, 1690 (1961).

13. A. J. Freeman and R. E. Watson, Phys. Rev. $\underline{127}$, 2058 (1962).

14. B. Bleaney, in "Proc. Third Int. Cong. Quantum Electronics", Paris, 1963, Ed. P. Grivet and N. Bloembergen (Columbia Univ. Press, New York, 1964).

15. B. R. Judd, Proc. Phys. Soc. $\underline{82}$, 874 (1963).

16. R. E. Watson and A. J. Freeman, in "Hyperfine Interactions", Ed. A. J. Freeman and R. B. Frankel (Academic Press, New York, 1967).

17. G. K. Woodgate, Proc. Roy. Soc. (London) $\underline{A293}$, 117 (1966).

18. A. Rosén, J. Phys. $\underline{B2}$, 1257 (1969).

19. S. K. Chan and D. J. Lam, in "Proc. 4th Int. Conf. Plutonium and Other Actinides", Santa Fe, New Mexico, 1970, Ed. W. N. Miner (Mett. Soc. Am. Inst. Mining, New York, 1970); G. M. Kalvius, M. B. Brodsky, B. D. Dunlap, Moshe Kuznietz, D. J. Lam, S. L. Ruby, and G. K. Shenoy, Proc. Int. Conf. Magnetism, Grenoble, 1970, in press.

20. N. Edelstein and R. Mehlhorn, Phys. Rev. $\underline{B2}$, 1225 (1970).

21. C. J. Lenander, Phys. Rev. $\underline{130}$, 1033 (1963).

22. L. Evans, P. G. H. Sandars and G. K. Woodgate, Proc. Roy. Soc. (London) $\underline{A289}$, 114 (1965).

23. Lloyd Armstrong, Jr. and Richard Marrus, Phys. Rev. $\underline{144}$, 994 (1966).

24. J. Bauche and B. R. Judd, Proc. Phys. Soc. (London) $\underline{83}$, 145 (1964).

25. For example, see G. J. Ehnholm, T. E. Katila, O. V. Lounasmaa, P. Reivari, G. M. Kalvius and G. K. Shenoy, Z. Physik, in press.

26. W. C. Easley, UCRL-17699, 1967.

27. R. Winkler, Z. Physik $\underline{184}$, 433 (1965).

28. A. J. Freeman, in "Hyperfine Structure and Nuclear Radiations", Ed. E. Matthias and D. Shirley (North-Holland, Amsterdam, 1968).

29. S. L. Ruby, G. M. Kalvius, B. D. Dunlap, G. K. Shenoy, D. Cohen, M. B. Brodsky and D. J. Lam, Phys. Rev. $\underline{184}$, 374 (1969).

30. For example, see S. Ofer, I. Nowik and S. G. Cohen, in "Chemical Applications of Mössbauer Spectroscopy", Ed. V. I. Goldanskii and R. H. Herber (Academic Press, New York, 1968).

31. K. W. H. Stevens, Proc. Phys. Soc. (London) $\underline{A65}$, 311 (1954).

32. L. Evans, P. G. H. Sandars and G. K. Woodgate, Proc. Roy. Soc. (London) A289, 114 (1965).
33. M. A. Coulthard, Proc. Phys. Soc. (London) 91, 44 (1967).
34. L. Evans, P. G. H. Sandars and G. K. Woodgate, Proc. Roy. Soc. (London) A289, 108 (1965).
35. H. Kopfermann, "Nuclear Moments", trans. E. Schneider (Academic, New York, 1958).
36. H. B. G. Casimir, "On the Interaction Between Atomic Nuclei and Electrons", (Teyler's Tweede Genootschap, Haarlem, 1936).
37. W. J. Childs, to be published.
38. J. B. Mann, Atomic Structure Calculations, Vol. I., Los Alamos Scientific Laboratory Report LA-3691, February 1968.

H_c-QS-IS CORRELATIONS IN OCTAHEDRAL IRON COMPOUNDS*

Y. Hazony

School of Engineering and Applied Science
and Computer Center, Princeton University,
Princeton, N. J., 08540

INTRODUCTION

While Mossbauer studies of particular iron compounds
have been very useful in illucidating theoretical questions,
by determining the symmetry or the character of the electronic ground state, its contribution to the overall understanding of chemical bonding in iron compounds has been rather
limited. Present theories of bonding in the first transition metal group have been developed mainly with the results
of optical spectroscopy in mind, thus predicting energies of
the electronic configurations. Mossbauer data, on the other
hand, reflect the charge density distribution (isomer shift,
IS, and quadrupole splitting, QS) or the details of the unpaired spin density (magnetic hyperfine interaction). Energy
calculations are quite insensitive to significant modifications of the electronic charge density distribution in solids
and it is not surprising that the results of such calculations do not contribute much to the interpretation of the
Mossbauer data. Furthermore, bonding theories have been
developed for the cases of extremely ionic and extremely covalent compounds and no single theoretical model is available to cover the entire range of covalency and to be used
for the interpretation of the experimental results. The
situation appears to be such that theories which have been
developed to account for the energetics of electronic levels
have more or less exhausted the information available from
optical spectroscopy and are awaiting for a different type

* Work supported in part by the U.S. A.E.C.

147

of experimental information for further confirmation and re-
finement or possible rejection. Apart from energy calculations, good theoretical predictions are available for the
free-ion unpaired spin density distribution, as reflected by
the magnetic hyperfine interaction at the nucleous [1].
Some attempts have been published to allow for covalency
effects, as reflected in the neutron magnetic form factor
data [2]. A phenomenological analysis of the available experimental results for octahedral divalent iron compounds
yields an empirical model which provides some guidance for
futher theoretical investigation [3]. This work will be
summarized here and extended to include low-spin iron compounds.

QS - IS CORRELATIONS

     Early attempts at systematic analysis of the QS-IS re-
lationship in iron compounds demonstrated that the situation
is rather complicated. This complexity is illustrated by
the commonly used expression for QS for high spin ferrous
compounds,

$$\Delta E = (2/7)e^2 Q(1-R)\alpha_c^2 <r^{-3}>_o F(\Delta_1, \Delta_2, \alpha_s^2 \lambda_o, T) + \qquad (1)$$

+ lattice contribution

with the usually accepted notation. It is not clear apriori
what is the relative importance of the different variables
involved in Eqn. 1. From the simplicity of the typical curves of the temperature dependence of ΔE (Fig. 1,[4,5]) it is
clear that such curves cannot possibly be used for a unique
determination of all the parameters involved. One may hope
to isolate some study-cases, where there are reasons to believe that the effects of some of the parameters in Eqn. 1
are negligible. For example, if the distortion from octahedral symmetry is large enough so that the separation of the
electronic ground state from the upper two levels (Δ_1, Δ_2)
is significantly larger than the spin-orbit interactions
($\alpha_s^2 \lambda_o$), and on the other hand the distortion is small enough
so that the lattice contribution to ΔE is relatively small,

$$\Delta E \simeq (\Delta E)_o \alpha_c^2 F \qquad (2)$$

would be a reasonably good approximation to Eqn. 1 at T = 0 K.
Possible variation of the antishielding factor (1-R) with
covalency are included in α_c^2. If those variations are be-

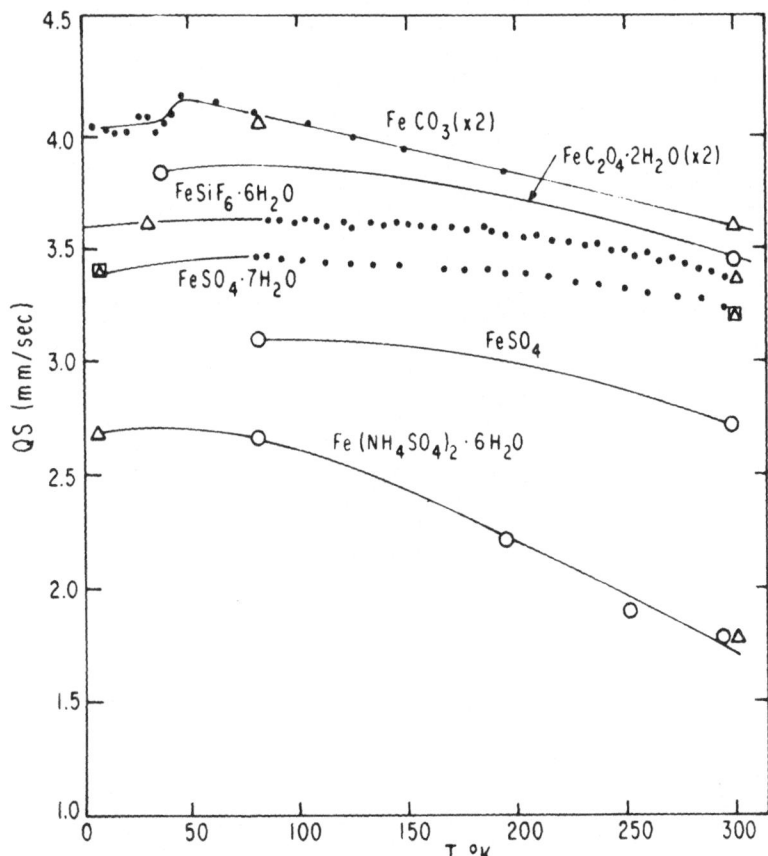

Figure 1. Typical quadrupole splitting data [4,5].
The results for FeCO$_3$ and FeC$_2$O$_4$·2H$_2$O are doubled (see text).
The low temperature anomaly in FeCO$_3$ is associated with a
metamagnetic phase transition (see text). The detailed data
for FeSiF$_6$·6H$_2$O and FeSO$_4$·7H$_2$O are from unpublished results
of Ref. 4.

lieved to be negligible then $\alpha_c^2 = \langle r^{-3}\rangle / \langle r^{-3}\rangle_o$ may be considered as a measure of the changes in the charge density
distribution of the t$_{2g}$ electron due to effects of covalency.
In the case of a doubly degenerate ground state, with $\Delta_1 = 0$
and Δ_2 significantly larger than the spin-orbit interaction,
the magnitude of ΔE will be reduced by a factor of F = 1/2.
For comparison with compounds with singlet ground state (F = 1)
one may use $\Delta E' = 2\Delta E$ to represent the compounds with a

doubly degenerate ground state. In order to further simplify
the analysis it seems reasonable to confine such studies to
circumscribed chemical series in which it is possible to establish simple relationships between the several compounds
involved. We have published such studies for three series of
high-spin octahedral compounds: FeX_2 (X = F, Cl, Br and I)
(see Fig. 2), the hydrates of $FeCl_2$ ($[FeCl_{6-n} \cdot nH_2O]^{n-4}$ with

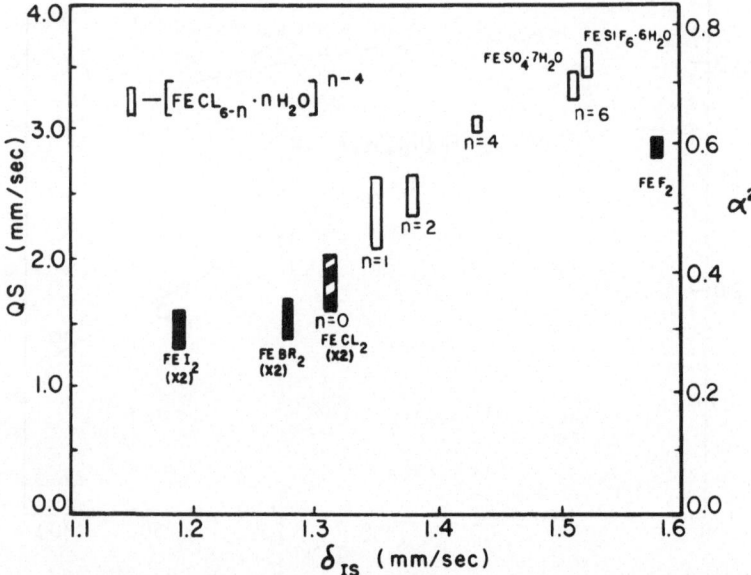

Figure 2. Quadrupole splitting versus isomer shift for
the ferrous halides (black bars) and the hydrates of $FeCl_2$
(clear bars). The top of the bars represents QS(90°K) and
the bottom QS(300°K). The width of the bars represent the
probable errors in δ_{IS}, which are the isomer shifts correct-
ed for the second order Doppler shift and the zero-point
motion [4] (versus metallic iron at room temperature). QS
data for $FeCl_2$, $FeBr_2$ and FeI_2 are doubled as discussed in
the text.

n = 0, 1, 2, 4, 6, see Fig. 2) and the hydrates of
FeI_2 ($[FeI_{6-n} \cdot nH_2O]^{n-4}$ with n = 0, 4 and 6, see Fig. 3) [4].
The compounds $FeCl_2$, $FeBr_2$ and FeI_2 are of trigonal symmetry
with a doublet electronic ground state at the paramagnetic
phase. At low temperatures they undergo a metamagnetic phase

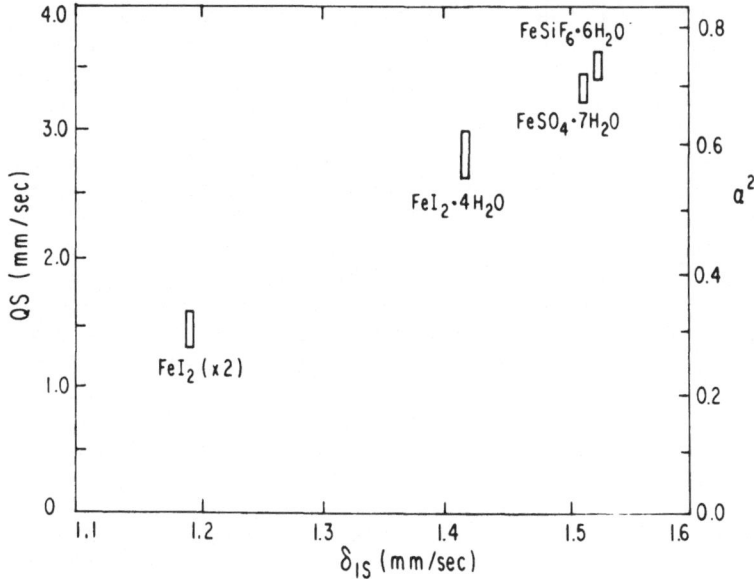

Figure 3. Quadrupole splitting versus isomer shift for the hydrates of FeI$_2$ [4] (see caption for Figure 2).

transition which may or may not be associated with structural deformations. This may remove the degeneracy of the electronic ground state and cause the increase in ΔE below these transitions. A comparison with other compounds of similar symmetry indicates that the ΔE results above the metamagnetic transitions better characterize the chemical bonds [3]. The question of the origin of the large increase in ΔE below the metamagnetic transitions in the above three compounds has not been settled experimentally. However, the large thermal hysteresis experienced in FeCl$_2$, upon lowering the temperature below T$_c$, may be considered as strong evidence that lattice deformations indeed occur [6]. We choose the above approach because, in contrast with the opposite one, it leads to a simple systematic representation of the Mossbauer data. The QS-IS relationships for the above three series are showing nearly linear correlations (Figs. 2 & 3). Such linear correlations indicate that, at least for these series of compounds, Eqn. 2 may be considered as a reasonable approximation to Eqn. 1 [3,4].

We have proposed that the surprisingly linear correla-

tions result from the similar effects on both QS and IS that
are produced by differences in the central field covalency
i.e., by differences in the radial portion of the iron atom's
3d electronic wave function, due to the modification of the
ionic effective charge on the ferrous ion [8]. This conclusion finds support in a careful analysis of $\Delta E(T)$ data in
$FeCl_2$ [7] and in comparison of Mossbauer hyperfine interactions (QS, IS, and magnetic) with data from crystal structure, neutron scattering, optical spectroscopy, ESR and EPR
studies [3]. These conclusions have been reinforced by the
theoretical results of Freeman and Ellis who constructed
radial 3d wave functions with variational rather than fixed
parameters [2]. Their fully variational, unrestricted HF-
SCF calculations then disclose that the t_{2g} wave functions
for $[MnF_6]^{4-}$ cluster are better described by an overall radial expansion rather than by an admixture with ligand orbitals.

Our work has emphasized the importance of separate discussions of circumscribed series of chemical compounds. The
simplicity of the correlations disappears when data from
different series are superimposed. This effect has been
attributed to the outstanding difference between the effects
of covalency on the $3d(t_{2g})$ and $3d(e_g)$ electronic wave functions [3,4].

Figure 4 extends the generality of the QS-IS correlations
discussed above [9]. The points of departure include two
additional groups of materials: (A) compounds with octahedrally co-ordinated oxygen atoms which participate in strongly
ionic complexes; and (B) low-spin covalent compounds of Fe^{III}.
QS vanishes for octahedral Fe^{II} covalent compounds because
of the low-spin configuration. One can, however, use the
data for the Fe^{III} complexes to estimate the contribution of
a single 3d electron to QS. No corrections have been made
for the IS values since they vary very little between the
Fe^{II} and Fe^{III} complexes of the same ligand configurations.
A third series (C) forms a bridge between the first two groups
and includes some of the compounds discussed previously plus
$FeTiO_3$ and FeS.

The right side of Fig. 4 indicates that IS saturates for
highly ionic compounds (group A). the vertical box contains
the set of compounds with saturated IS. A contrasting situation obtains for the covalent compounds (group B) which
appear on the left side of the figure. In these, covalency

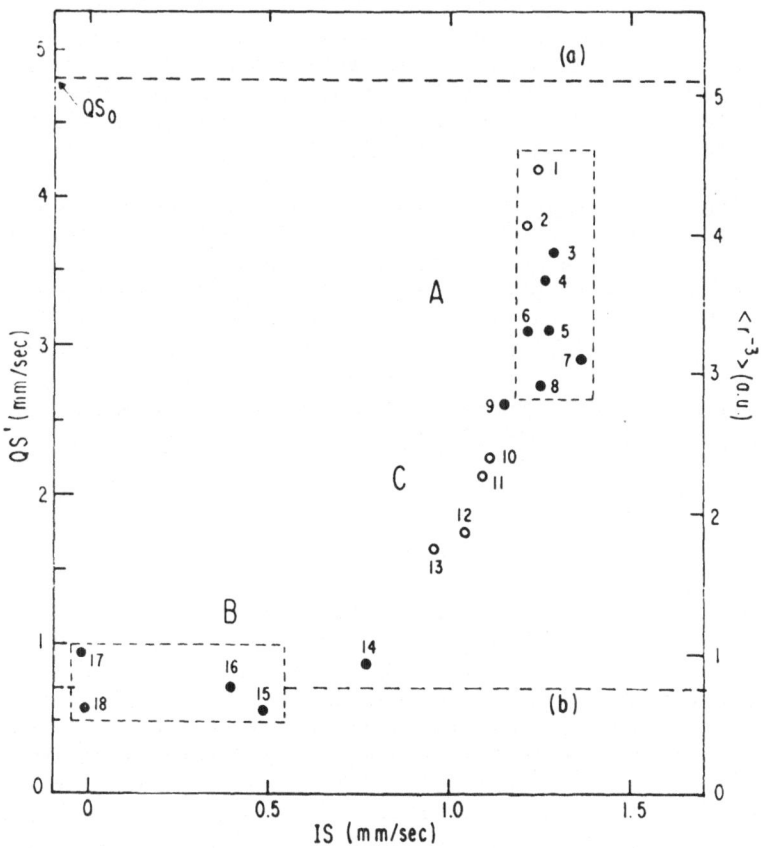

Figure 4. QS-IS relationship in octahedral iron compounds [9]. Full circles: QS' = QS; open circles: QS' = 2QS (see text). IS values are uncorrected room temperature data versus metallic iron. In the heavier halides ΔE data immediately above the transitions are considered. 1.FeCO$_3$, 2.FeC$_2$O$_4$·2H$_2$O, 3.FeSiF$_6$.6H$_2$O, 4.FeSO$_4$.7H$_2$O, 5.FeSO$_4$, 6.FeCl$_2$.4H$_2$O , 7.FeF$_2$, 8.Fe(NH$_4$SO$_4$)$_2$.6H$_2$O, 9.FeCl$_2$.2H$_2$O, 10.FeTiO$_3$, 11.FeCl$_2$,12.FeBr$_2$, 13.FeI$_2$, 14.FeS, 15.FeS$_2$ (derived from the differences in QS between FeS$_2$, FeSAs and FeAs$_2$[3]), 16.Fe^{III} N,N-Dialkyldithiocarbamates, 17-18.Fe^{III}(CN)$_6$$^{3-}$ complexes.

reduces the effective charge, Z^*, on the iron atom to a
negligible magnitude. Since the radial portion of the 3d
wave function depends strongly on Z^*, the mean 3d density
reaches a minimum at the "free-atom" value. The reduction
of Z^* may be achieved within a range of S to d density ratios,
hence large variations of IS may be expected and, indeed, are
observed. This IS range coincides with that of metallic
iron and the iron alloys for which $Z^* \approx 0$ is also expected.

The wide variation of QS for ionic compounds whose IS
has saturated (points 1-9, Fig. 4) follows because QS depends
primarily upon the $3d(t_{2g})$ electron density while, as pointed
out above, IS reflects the total 3d and s electron densities.
The dashed horizontal line (a) toward the top of Fig. 4 denotes $\Delta E_Q = 4.8$ mm/sec, which corresponds to a hypothetical,
completely "ionic" Fe^{2+} compound. The magnitude of ΔE_Q derives from the calibration of the covalency parameter and
corresponds to $\alpha_c^2 = 1$ [3,4].

The lower dashed line (b) in the figure represents an
estimated limit for $<r^{-3}>$ for a single d electron in covalent or atomic iron (i.e., effective charge $Z^* \approx 0$). This
limit (~ 0.7 mm/sec or $<r^{-3}> \sim 0.75$ a.u.) which is deduced
from the data for the ferricyanide complexes, implies a
large expansion of the t_{2g} wave function. This is consistent with the recent observation based on molecular dynamical
analysis, that the t_{2g} electrons in iron-cyanides primarily
affect the inter- rather than intra-molecular forces. Similar
values may be derived from the data for $Fe^{III}N$, N-Dialkyl-
dithiocarbamates, or from the analysis of the data for the
series FeS_2, FeSAs and $FeAS_2$ with t_{2g}^6, t_{2g}^5 and t_{2g}^4 configurations, respectively [3] . The fact that the removal of one
3d electron has a very small effect on the IS of the above
covalent compounds is consistent with the notion of large
radial expansion of the 3d orbitals.

Data for $FeTiO_3$ and FeS fall smoothly on the linear
curve established earlier for the ferrous halides and indicate, perhaps not too surprisingly, that the transition from
"ionic" to "covalent" is continuous, insofar as the charge
density distribution is concerned.

MAGNETIC HYPERFINE INTERACTION

The observed internal magnetic field at the nucleus may
be described in terms of three components,

$$H_{int} = H_c + H_L + H_D \tag{3}$$

H_c is the Fermi contact interaction caused by the interaction of the nuclear magnetic moment with the spin density distributions of the atomic s-electrons at the nucleus, polarized via exchange interaction with the unpaired d electrons. H_L is the contribution from the orbital current due to the unquenched angular momentum. H_D is the component due to the anisotropy of the spin density distribution and should vanish for a strictly cubic ligand configuration. Eqn. 3 may be rewritten as [10]

$$H_{int} = H_c + 4\beta<r^{-3}> \Omega_i \tag{4}$$

where

$$\Omega_i = \frac{1}{2} (g_i-2)S + \frac{1}{14} <0|L_i^2 - 2|0>$$

g_i and L_i are the components of the electronic g-factor and the angular momentum operator respectively. While the second term in Eqn. 4 depends linearly on $<r^{-3}>$ of the single t_{2g} electron, the contribution of the contact term, H_c, depends on both unpaired t_{2g} and e_g electrons.

   The problem of theoretical interpretation of QS(T) and H_{int} data for a single compound involves too many unknown parameters and is underdetermined. It would be appropriate therefore to apply the analysis simultaneously to a series of compounds. In our attempt to normalize the $<r^{-3}>$ scale for the t_{2g} electron we will compare the H_{int} data for a group of compounds ($FeCl_2.2H_2O$, FeF_2, $Fe_3(PO_4)_2.8H_2O$ and $FeCl_2.4H_2O$). These compounds have similar ΔE values at low temperatures (2.6, 2.9, 3.0 and 3.1 mm/sec. respectively) and would be expected to be associated with $<r^{-3}>$ of similar magnitudes.

   The analysis of H_{int} for these compounds in terms of Eqn. 4 is shown in Fig. 5 and the relevant parameters are tabulated in Table 1. These data have been first discussed by Johnson who observed the approximate linear relation between H_{int} and Ω_i. He observed that this is consistent with single values for $<r^{-3}>$ as well as H_c for this group of compounds [10]. In our previous discussions of the QS-IS correlations in the ferrous halides and their hydrates we

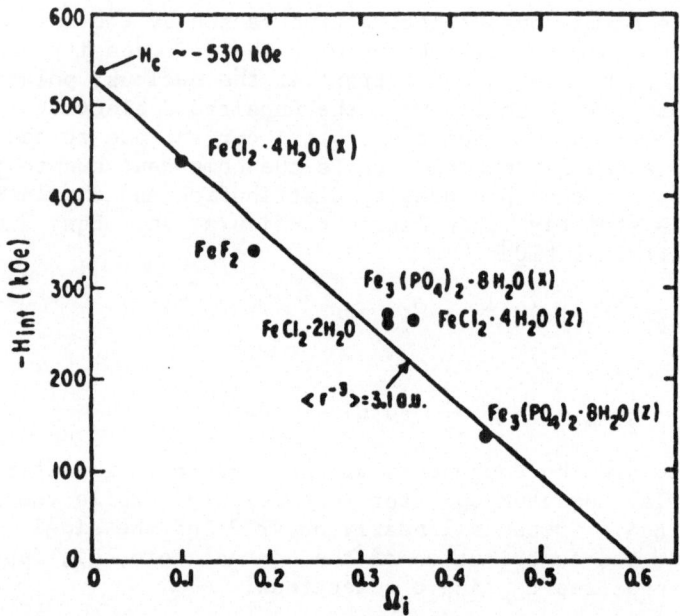

Figure 5. H_c vs Ω_i for some Fe^{2+} compounds [3].

TABLE 1

	ΔE ($\frac{mm}{sec}$)	i	H_i (kOe)	Ω_i
$FeCl_2 \cdot 4H_2O$	3.1	x	-440	0.10
		z	-266	0.36
$Fe_3(PO_4)_2 \cdot 8H_2O$	3.0	x	-270	(0.33)
		z	-135	(0.44)
FeF_2	2.9	x	-330	0.18
$FeCl_2 \cdot 2H_2O$	2.6	x	-260	0.33

These data are taken from ref. 3 and references there in, ΔE are the experimental values at low temperatures.

have assigned $\alpha_c^2 \approx 0.6$ for FeF_2. This value, multiplied by the theoretical estimate of $<r^{-3}>_o = 5.1$ a.u., yields $<r^{-3}> \approx 3.1$ a.u. for FeF_2. In view of the similarity in ΔE values observed for the above compounds (Table 1), the straight line associated with the $<r^{-3}> = 3.1$ a.u. is expected also to fit the H_{int} data, which is indeed the case (Fig. 5). The internal consistency between the QS-IS analysis and the results of the internal magnetic fields is reassuring. The good agreement between the predictions of the two methods indicates that $<r^{-3}>$ values measured by both of them are closely the same in cases of moderate covalency. Consequently, one can use the predictions from QS data in the analysis of the H_{int} results.

The magnitude of the contact interaction term, H_c, is obtained from the intersection of the straight line in Fig.5 with the $\Omega_i = 0$ ordinate. The value of $H_c = -530 \pm 40$ kOe so obtained is very close to the calculated one of -550 for the Fe^{2+} free ion [1]. The large radial expansion of the t_{2g} orbitals in FeF_2, as indicated by the rather small α_c^2 value, has practically no effect on H_c. This is implying that the average unpaired 3d spin density, at the central portion of the metal ion, is very close to that of the free ion. The average unpaired spin density is the sum of two contributions and the apparent paradox must be due to a mutual cancellation of covalency effects on the two t_{2g} and two e_g electrons [3]. In other words, the 40 per cent depletion of the t_{2g} spin density in FeF_2 is compensated for by a similar increase of 40 percent in the e_g spin density. The ratio between the e_g and t_{2g} spin densities at the central portion of the ferrous ion in FeF_2 is given by $(1.4/0.6) \approx 2.3$.

Further support for this effect may be derived from the radial scaling factors available for fluorides and oxides of Mn^{2+} (1.11), Co^{2+} (0.87) and Ni^{2+} (0.83) [11-13]. A radial scaling factor of ~ 1.0 is obtained for FeF_2 by interpolation between these results (Fig. 6) which is consistent with the notion of mutual cancelation of covalency effects.

The effective volume occupied by the contracted 3d distribution in NiO is given by the cube of the observed scaling factor of 0.83. Allowing for 18 per cent charge transfer to the ligands [14], a 45 percent increase in the mean 3d spin density is obtained for NiO. The agreement between the estimates for the e_g contraction in FeF_2 and NiO is significant.

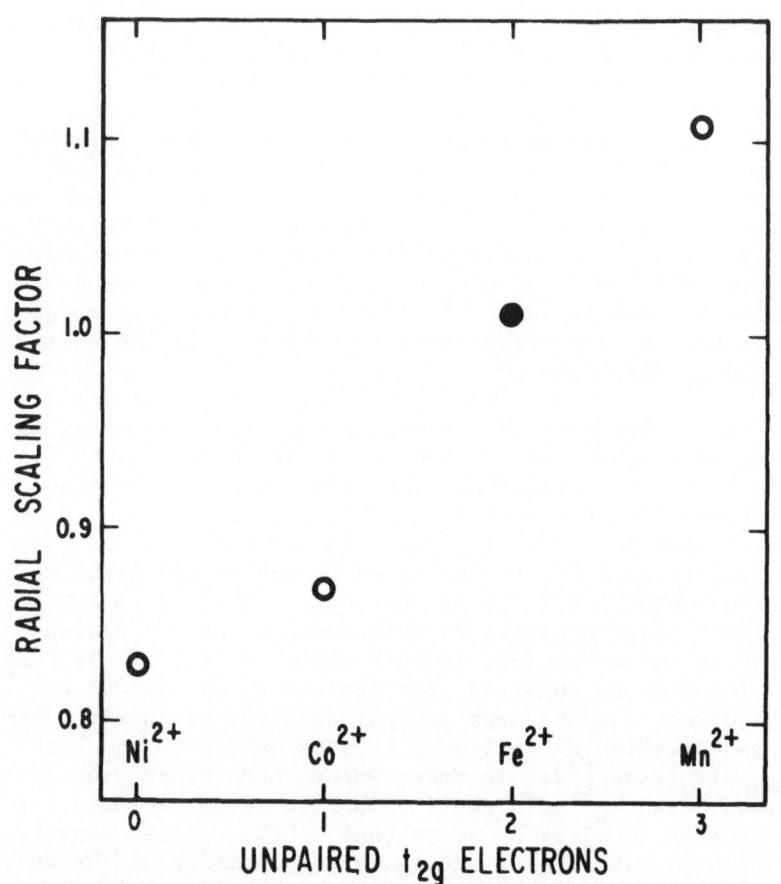

Figure 6. Radial scaling factors <u>vs</u> number of unpaired
t_{2g} electrons. Open circles represent neutron scattering
data while full circle is derived from the H_c data for FeF_2.

The extent of radial expansion for the $3d(t_{2g})$ electrons
in $(MnF_6)^{4-}$ may be estimated from the theoretical results of
Freeman and Ellis [2]. A comparison between the neutron
scattering form factors for the t_{2g} component of the unpaired
spin density and the corresponding free-ion results (Figs.
1 and 2 in ref. 2) shows that the difference between them
could be described to a first approximation by a radial scaling factor of 1.15. The corresponding reduction in $<r^{-3}(t_{2g})>$,
as compared with the free-ion value, is given by $1.15^{-3}=0.66$.
This is in good agreement with the value of ~ 0.6 for FeF_2,
particularly in view of the rather limited radial flexibility

of the minimal basis set used in these theoretical computations [2].

H_c-QS CORRELATION

In view of the consistency observed above between the predictions for $<r^{-3}>$ of the t_{2g} electron, from QS and H_{int} analysis, it would be instructive to extend the comparison to the other members of the ferrous halides series. This can be done using the data for additional two compounds, $FeCO_3$ and $FeTiO_3$, which are of the same symmetry of ligand configuration as $FeCl_2$, $FeBr_2$ and FeI_2, and for which a detailed analysis has been published by Okiji and Kanamori [15]. We will apply the results of Okiji and Kanamori to all five compounds, using $<r^{-3}>$ estimates obtained from the ΔE data. Because of the double degeneracy of the t_{2g} ground level, we will use $\Delta E' = 2.\Delta E$ (above the corresponding metamagnetic transitions).

To test this approach we will estimate $<r^{-3}>$ for $FeCO_3$ by comparing it to FeF_2. Using $\Delta E' = 4.2$ mm/sec. for $FeCO_3$ and $\Delta E = 2.9$ mm/sec. and $<r^{-3}> = 3.1$ a.u. for FeF_2, the estimate would be $<r^{-3}> = 4.4$ a.u. for $FeCO_3$. This is the same value as used by Okiji and Kanomori in their analysis [15] and is significantly smaller than the "free-ion" value of 5.1 a.u., proposed by Watson and Freeman [1]. Assuming that the difference includes both effects of radial expansion and MO type mixing with ligand orbitals, the second term in Eqn.4 is estimated for $FeCO_3$ as $H_L + H_D \approx + 730$ kOe. Comparing with the experimental result of $H_{int} = + 184$ kOe [16] one obtains $H_c \approx -545$ kOe for $FeCO_3$. The same analysis is carried out for the heavier halides and $FeTiO_3$, using Eqn. 4 and $<r^{-3}>$ values derived from the ΔE data. The results are tabulated in Table 2 and displayed in Fig. 7 as a function of $<r^{-3}>$.

From the relation between H_c and $\Delta E'$ (Fig.7) and between IS and $\Delta E'$ (Fig. 4) it appears that both H_c and IS saturate toward the ionic end of the correlation curves. This is due to the mutual cancelation of e_g contraction and t_{2g} expansion for moderate degree of covalency. In the more ionic compounds, like FeF_2, the attraction, between the Fe^{2+} ion and its relatively small ligands, is balanced by the compression of the electronic cloud along the bond directions. Such a compression may account for the apparent contraction observed in NiO for the $3d(e_g)$ wave functions. Because of the increase of the ionic radii of the ligands in $FeCl_2$, $FeBr_2$ and FeI_2, the

TABLE 2

	ΔE ($\frac{mm}{sec}$)	$\Delta E'$ ($\frac{mm}{sec}$)	$<r^{-3}>$ (a.u.)	H_{int} (kOe)	H_L+H_D (kOe)	H_c (kOe)
$FeCO_3$	2.1	4.2	4.4	+184	+730	-546
$FeTiO_3$	1.14	2.28	2.4	-70	+400	-470
$FeCl_2$	1.03	2.06	2.15	+3	+358	-355
$FeBr_2$	0.86	1.72	1.8	+30	+300	-270
FeI_2	0.81	1.62	1.7	+74	+280	-206

These data are taken from ref. 3 and references therein. ΔE values are for $T \gtrsim T_c$. $\Delta E' = 2\Delta E$.

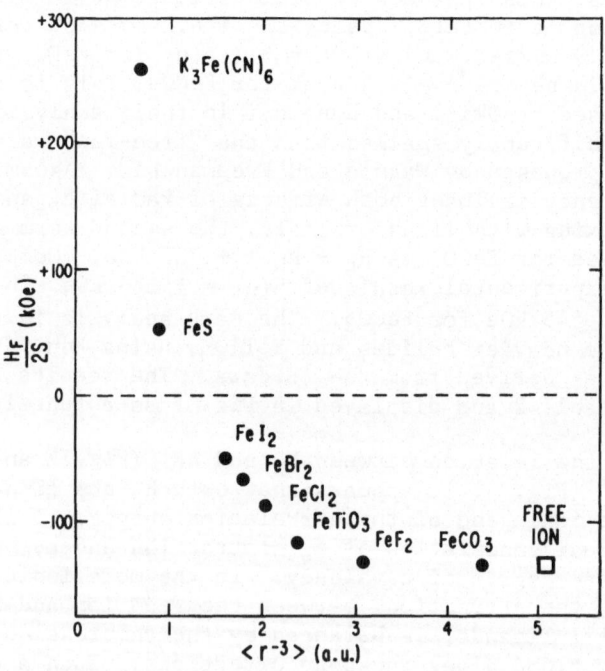

Figure 7. ($H_c/2S$) vs $<r^{-3}(t_{2g})>$ for some high-spin ferrous compounds [3] and the low-spin complex $K_3Fe(CN)_6$.

crystal structure is that of a closed packed arrangement of
the ligands, with the Fe^{2+} ions occupying a sublattice of
interstitial positions. The attraction between the Fe^{2+}
ions and the ligands is balanced to a large extent by the
overlap interaction between the ligands in the closed packed
lattice. This increase in anion-anion repulsion reduced the
compression of the charge density along the cation-anion bond
direction. The increase of the size of the interstitials
occupied by the Fe^{2+} ions reduces the contraction of the e_g
charge density as well as provides room for the expansion of
the t_{2g} wave functions. The net effect is that of expansion
of the total spin density distribution. The observed decrease
in magnitude of H_c is consistent with the theoretical results
of Watson and Freeman. From their work one could anticipate
a decrease in magnitude, with an eventual change of sign of
H_c, as a result of significant radial expansion of the 3d
wave functions [1].

     Within the framework of the present interpretation of
the H_c-QS relationship, in the octahedral high-spin ferrous
compounds, one may predict by extrapolation that any further
increase in covalency would result in a positive value for
H_c. An appropriate compound for an extension of the present
analysis is FeS. A comparison of the QS-IS values of FeS
with those for the compounds discussed above (Fig. 4) supports
such an extension. Because of the rather small $<r^{-3}>$ indicated by the small value of $\Delta E \approx 0.9$ mm/sec.[17], the correction due to the second term in Eqn. 4 is expected to be small.
This correction is estimated as $+ 100 \pm 50$ kOe and, depend-
ing on the sign, the measured value of $|H_{int}| = 309$ kOe [17]
corresponds to H_c values of $+ 210 \pm 50$ or -410 ± 50 kOe. The
present analysis (Fig.7) favors $H_c \sim +210 \pm 50$ kOe for FeS
and consequently the positive sign for H_{int}. The sign of
H_{int} for FeS has not been determined experimentally.

                    LOW-SPIN IRON COMPOUND

     Any further increase of covalency would cause a change
from the high-spin to a low-spin electronic configuration
and therefore to the vanishing of both QS and H_{int} in octahedral divalent iron complexes. One could explore, however,
the implications of the above discussions for low-spin compounds by considering the effects of the removal of a single
3d electron from an octahedral Fe^{II} configuration. The miss-
ing single 3d electron, from an otherwise complete t_{2g} subshell, results in the observation of both QS and H_{int} in low-

spin Fe^{III} compounds. Such data is available for $K_3Fe(CN)_6$ which displays at low temperatures ΔE = 0.52 mm/sec., [9], $H_{||}$ = +193 kOe and $H_{\perp}$= +253 kOe [18]. From Eqn. 4 and the available $|g_{||}|$ = 0.915 and $|g_{\perp}|$ = 2.22 [19] it is apparent that $H_{||}$ determines a lower limit for H_c in this case. A more detailed analysis yields an estimate of $(H_c/2S) \approx$ +250 kOe which is to be compared with $(H_c/2S) \approx$ -124 kOe in FeF_3 [20]. The corresponding data point is compared in Fig. 7 with the above results for the high-spin ferrous compounds. A more appropriate comparison would be with the high-spin trivalent compounds. However, because of the $3d^5$ electronic configuration, octahedral Fe^{3+} compounds would not show any quadrupole splitting (significant distortion from the octa-hedral symmetry would result in the observation of QS which is not directly related to the 3d density distribution). The approximately linear H_c-IS relationship, displayed in Fig. 8, could have been anticipated in view of the similarity between the overall QS-IS and QS-H_c correlations (Figs. 4 & 7). Here again a comment should be made with regard to the data point associated with $K_3Fe(CN)_6$. This Fe^{III} data point has been used since H_c vanishes for Fe^{II} complexes. The compari-son is instructive since in both cases we consider the con-tribution of a single spin unit to H_c. The IS value has not been corrected for the difference of +0.1 mm/sec. between the trivalent and divalent iron cyanides. Such a correction would improve the nearly linear H_c-IS relationship. The lin-ear H_c-IS correlation is significant in view of the fact that H_c depends on the details of the unpaired spin density dis-tribution while IS depends on the distribution of the total charge density. While LCAO-MO theory does not predict any obvious simple relation between the two observables, a simple relation would be expected if the radial scaling is the domin-ant effect of covalency.

SUMMARY

The main purpose of the present work is to present a model which describes the variation of the 3d charge density distribution with covalency. This model applies to the "ionic", "covalent" as well as intermediate iron compounds with approximately octahedral symmetry.

The main features of the model are:
a. A large range of variation of the mean $3d(t_{2g})$ charge density as reflected by a factor of $\sim$ 7 decrease in $<r^{-3}(t_{2g})>$ on going from the extremely ionic to the fully covalent com-pounds;

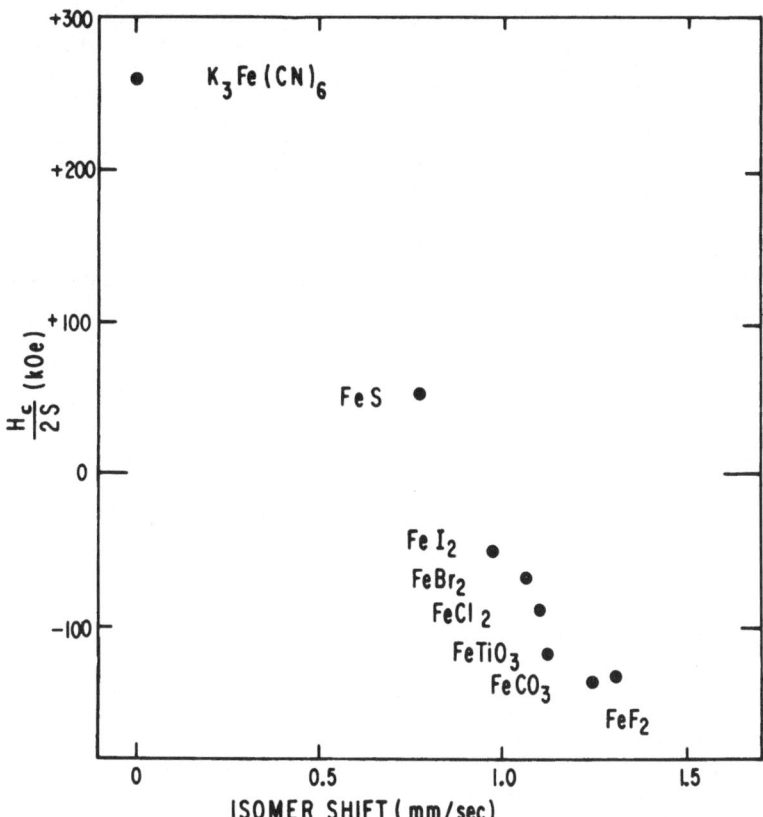

Figure 8.   (H$_C$/2S) vs IS with respect to metallic iron at room temperature for some high-spin ferrous compounds and the low-spin complex K$_3$Fe(CN)$_6$.

b.   A significant increase of the mean 3d(e$_g$) charge density with moderate increase of covalency. This effect accounts for the observation that the contact interaction term for FeF$_2$ is very close to the free-ion value, while $<r^{-3}(t_{2g})>$ is about 40 percent smaller than the free-ion value. For further increase of covalency the expansion effect dominates.

While this model has been developed mainly to account for the Mossbauer data, it has ample support from results of other experimental techniques. Its relation to the results of neutron scattering experiments has been discussed in the text. Further support may be derived from the results of crystal structure investigations as reflected by the crystal-

line radii of the ions. The crystalline radii provide a measure of the mean value <r> and indicate a change of about a
factor of two between the extremely ionic and fully covalent
cases [21]. Additional support may be derived from the data
available from optical spectroscopy, showing a change of
about a factor of two in the coulombic repulsion term B,
which is a measure of $<r^{-1}>$[8].

Comparing the three different averages of the radial
distribution of the 3d wave functions, $<r^{-3}>$, $<r^{-1}>$ and $<r>$
one finds

$$\left(\frac{<r^{-3}>_{cov}}{<r^{-3}>_{ion}} \right)^{-1/3} \cong 1.9 \qquad \text{(Mossbauer)}$$

$$\left(\frac{<r^{-1}>_{cov}}{<r^{-1}>_{ion}} \right)^{-1} \cong 2.1 \qquad \text{(Optical)}$$

$$\frac{<r>_{cov}}{<r>_{ion}} \cong 1.7 \qquad \text{(Crystal radii)}$$

Such a similarity between the different moments of the radial
wave functions would be expected if radial scaling is the
dominant effect of covalency.

Further evidence for the significant expansion of the
t_{2g} wave functions is available from the EPR data for FeF_2
[22] and NMR data for $KMnF_3$[23]. The results in both cases
indicate large transfer of t_{2g} spin densities toward the
spaces in between the ligands, which is consistent with the
large depletion of the t_{2g} densities at the central ion. It
appears that LCAO theory displays enough flexibility to
account for the expansion of the t_{2g} orbitals in the fluorides
of the transition metals. However it requires similarly large
magnitudes of t_{2g} and e_g interactions. These results are not
disturbing once it is recognized that they compensate for the
lack in flexibility in the radial wave functions. Within the
framework of the present model, the large radial expansion
would increase the amount of t_{2g} overlap and spin transfer
toward the spaces between the ligands; the accompanying effect
of e_g contraction would reduce the e_g overlap, qualitatively
accounting for the EPR results for FeF_2 and NMR data for

KMnF$_3$. A quantitative analysis must await more detailed fully-variational theoretical computations, possibly with more extended basis sets.

Finally, the fluorides are frequently used as model compounds for ionic calculations in solids, however, the systematics of Figs. 4 and 7 clearly indicates that FeF$_2$ is far from being an ideal study-case. Insofar as the concept of "ionic compounds" stands for the applicability of unperturbed free-ion 3d wave functions, FeCO$_3$ represents the smallest distortion of the 3d wave function. The fact that on the IS,H$_c$ and electronegativity scales FeF$_2$ appears to be a most ionic ferrous compound derives from the mutual cancelation of opposing large covalency effects on the two t$_{2g}$ and two e$_g$ 3d electrons. This effect allows to apply by extrapolation the saturation values of IS and H$_c$ (Figs. 4 & 7) in a direct comparison with the results of theoretical free-ion computations. This extrapolation confirms the theoretical result of H$_c$ = -550 kOe for the free-ion Fe^{2+} [1]. Similarly, one can estimate the free-ion value of $<r^{-3}>$ using the absolute value derived from the slope in Fig. 5 in conjunction with the independent calibration of the α_c^2 scale, based on ESR data [4]. This proceedure confirms the calculated result of $<r^{-3}>_0$ = 5.1 a.u. for the Fe^{2+} free-ion [1]. Based on this one can associate the saturation IS value of 1.3 mm/sec. (with respect to metallic iron at room temperature) with the calculated s-electron density distribution for the Fe^{2+}(3d^6) free-ion.

REFERENCES

1. R. E. Watson and A. J. Freeman, Phys. Rev. 123, 2027 (1961).
2. A. J. Freeman and D. E. Ellis, Phys. Rev. Letters, 24, 516 (1970).
3. Y. Hazony, Phys. Rev. B3, 711 (1971).
4. Y. Hazony, R. C. Axtmann and J. W. Hurley, Jr., Chem. Phys. Letters, 2, 440 (1968); ibid, 2, 673 (1968); J. Chem. Phys. 52, 3309 (1970).
5. H. N. Ok, Phys. Rev. 185, 472 (1969); R. W. Grant, H. Wiedersich, A. H. Muir, Jr., U. Gonser and W. N. Delgass, J. Chem. Phys. 45, 1015 (1966); Deo Raj, K. Chandra and S. P. Puri, J. Phys. Soc. Japan, 24, 35 (1968).
6. Footnote 21 in Ref. 7.
7. Y. Hazony and H. N. Ok, Phys. Rev. 188, 591 (1969).
8. C. K. Jørgensen, Oxidation Numbers and Oxidation States, Springer-Verlag, New York, 1969.

9.   Y. Hazony and R. C. Axtmann, Chem. Phys. Letters $\underline{8}$,
     571 (1971).
10. C. E. Johnson, Symp. Faraday Soc. $\underline{1}$, 7 (1967).
11. H. A. Alperin, Phys. Rev. Letters, $\underline{6}$, 55 (1961).
12. D. C. Khan and R. H. Erickson, Phys. Rev. $\underline{B1}$, 2243 (1970).
13. J. M. Hastings, N. Elliott and L. M. Corliss, Phys. Rev.
     $\underline{115}$, 13 (1959).
14. J. Hubbard and W. Marshall, Proc. Phys. Soc. $\underline{86}$, 561
     (1965).
15. A. Okiji and J. Kanamori, J. Phys. Soc. Japan, $\underline{19}$, 908
     (1964).
16. D. W. Forester and N. C. Koon, J. Appl. Phys. $\underline{40}$, 1316
     (1969).
17. S. Hafner and M. Kalvius, Zeit. fur Krist. $\underline{123}$, 443
     (1966).
18. K. Ono, M. Shinohara, A. Ito, N. Sakai and M. Suenaga,
     Phys. Rev. Letters $\underline{24}$, 770 (1970).
19. J. M. Baker, B. Bleaney and K. D. Bowers, Proc. Phys.
     Soc. (London) $\underline{B69}$, 1205 (1956).
20. G. K. Wertheim, H. J. Gugenheim and D.N.E. Buchanan,
     Phys. Rev. $\underline{169}$, 465 (1968).
21. L. Pauling, The Nature of the Chemical Bond, third ed.,
     Cornell University Press, 1960.
22. M. Tinkham, Proc. Roy. Soc., $\underline{A236}$, 549 (1956).
23. R. G. Shulman and K. Knox, Phys. Rev. Letters, $\underline{4}$, 603
     (1960).

THE MÖSSBAUER PARAMETERS IN TIN(IV) AND EUROPIUM(III) COMPOUNDS

A.J. Carty and H.D. Sharma*

Department of Chemistry
University of Waterloo
Waterloo, Ontario, Canada

There is a considerable interest in the interpretation of the Mössbauer parameters of Sn(IV) and organotin (IV) compounds. A compilation of the isomer shift, δ, and the quadrupole splitting, ΔEq, data on organotin compounds (1) and reviews on this topic have appeared (2-4). Tin-119 isomer shifts reflect changes in 5 s electron density at the nucleus. It has been shown that the contribution of inner shells remains virtually constant for both emitters and absorbers in Sn(IV) compounds (3). The δ value is known to be dependent on populations of Sn(IV) ion valence states i.e. s, p and d orbitals relative to those of the emitter (5). Expressions relating isomer shifts with the electronic structure of tin in its compounds have been obtained by Goldanskii et al (6) by following Fermi-Segré formulae for the electron density of 5 s electrons. Ruby et al obtained almost linear relationships of δ with the electron populations of valence states in tin for different numbers of s and p electrons using Hartree-Fock self consistent-field atomic wave functions (7). Equations relating δ with the population of tin ion states have the general form:-

$$\delta = C[s - C'p - C''p^2] + C''' \qquad [1]$$

where s and p are the population densities of s and p orbitals of the tin atom C, C', C'' and C''' are constants.

The experimental results of Herber and Cheng (8), Ali et al (9), Clausen and Good (10) and Davies et al (11)

indicate that in compounds SnX_6^{2-} $SnX_4Y_2^{2-}$, $SnX_2(ox)_2$ and
$SnX_4 (oxH)_2$ (X and Y = halogen and oxH = 8 hydroxy quinoline), the isomer shift shows a linear relationship with
respect to a revised scale of average Mulliken electronegativities of halogen atoms forming Sn-X bonds. A similar
relationship is indicated for SnX_4 compounds (10,12). However the isomer shifts of organotin (IV) compounds do not
follow the above trends (12). It has been suggested that in
compounds in which there is a large disparity between the
polarities of the various tin-ligand/anion bonds, other
factors like bond polarity, number and disposition of the
tin-ligand/anion bonds become operative. Recent measurements of Mossbauer parameters on adducts of substituted
phosphines with $SnCl_4$ indicate that there is no preferential
concentration of s character in σ bonds with more electro-
positive ligands (13). It is the purpose of this article
to present alternative suggestions for relating δ and ΔEq
to bonding parameters in Sn(IV) and organotin (IV) compounds.
We also present some of our results on Eu(III) compounds.

$L_n SnX_{6-n}$ Compounds. It has been shown that in a series
of hexahalo and mixed hexahalo compounds, δ values show a
better linear relationship with the revised scale of
Mulliken electronegativity rather than with Pauling electronegativities of halogen atoms (10-12). Without entering
into any controversy over the merits and demerits of the
various proposed scales of electronegativity, we adopt the
former scale. From the computations of Ruby et al, it
appears that $|\psi(0)|^2_s$ of Sn(IV) is directly related to the
electronegativities of ligands/anions forming bonds in the
compounds. The population of s, p and d orbitals in hybrid
orbitals of tin atoms bears a direct relationship to the
electronegativity of ligands. An expression for δ is derived from the experimental values obtained for hexahalo
compounds

$$\delta = -0.0717 \ [n\bar{X}_A + \ldots m\bar{X}_B + \ldots(6\text{-}n\text{-}m\ldots) \ X_i] + 4.13^* \quad [2]$$

where δ = isomer shift in $mm.sec^{-1}$ with respect to a SnO_2
source. n,m = number of ligands or anions A and B forming σ
bonds with the tin atom. $\bar{X}_A, \bar{X}_B..\bar{X}_C$ = Mullikan electronegativities of ligands/anions A,B,C,....i. ($\bar{X}..$ F = 10.43, Cl =
8.29, Br = 7.59 and I = 6.75).
* This is a simplified equation and is applicable over a limited range of δ values. A complete equation will be
  presented elsewhere.

The calculated values of $\bar{X}$ of ligands/anions for a few representative compounds are given in Table 1. The reported δ value for $[(C_6H_5)_3P]_2 SnBr_4$ is in error (14). Measurements on this compound in our laboratory give $\delta = 0.95$ mm.sec^{-1} which is in agreement with the one derived from the δ value of $[(C_6H_5)_3P]_2 SnCl_4$. The molecular electronegativity of the NO_3^- ion obtained from the δ value of $Sn(NO_3)_6^{2-}$ is 9.28 which agrees with the value of 9.35 obtained through n.m.r. measurements (43). A few comments regarding trends in the calculated values of electronegativities are worth stating, namely a) the molecular electronegativities of ligands/anions are within ± 0.7 of the electronegativity of the donor atom which is consistent with the observation that the nearest neighbours of the Mossbauer atom in a molecule determine its Mossbauer parameters and electronic structure whilst the more remote atoms usually provide a much weaker effect which may be regarded as a perturbation. b) The electronegativity of the donor atom follows the following order

Table 1

Calculated Electronegativities of Ligands
in $L_n SnX_{6-n}$ compounds

Donor Atom Oxygen

Compound	δ mm.sec^{-1}	ΔE_q mm.sec^{-1}	$\bar{X}_L$
$[(CH_3)_2SO]_2SnCl_4$	0.38		9.57
$[(C_2H_5)_2SO]_2SnCl_4$	0.36		9.71
$[(n-C_3H_7)_2SO]_2SnCl_4$	0.38	0.59	9.57
$[(n-C_4H_9)_2SO]_2SnCl_4$	0.37	0.67	9.64
$[(CH_2)_4SO]_2SnCl_4$	0.34		9.85
$[(CH_3)_2SO_2]SnCl_4$	0.38	0.916	9.57
$[(C_2H_5)_2SO_2]SnCl_4$	0.406	0.863	9.39
$[(n-C_3H_7)_2SO_2]SnCl_4$	0.433	1.571	9.20
$[(n-C_4H_9)_2SO_2]SnCl_4$	0.393	0.830	9.48
$[(C_6H_5)_2SO_2]SnCl_4$	0.426	1.072	9.25
$[(CH_2)_4SO_2]SnCl_4$	0.508	1.150	8.68*
$(Cl_3PO)_2SnCl_4$	0.51	1.12	8.67†
$[Cl_2(C_6H_5)PO]_2SnCl_4$	0.39	1.61	9.50
$[(C_6H_5)_3PO]_2SnCl_4$	0.27	0.51	10.34
$[(n-C_4H_9)_3PO]_2SnCl_4$	0.24		10.55
$[(C_6H_5O)_3PO]_2SnCl_4$	0.34	0.71	9.85
$[Cl(C_6H_5O)_2PO]_2SnCl_4$	0.38	0.75	9.57

Compound	δ mm.sec^{-1}	ΔE_q mm.sec^{-1}	$\bar{X}_L$
$[Cl_2(C_6H_5O)PO]_2SnCl_4$	0.42	1.13	9.29
$[(C_6H_5)_3AsO]_2SnCl_4$	0.437	0.70	9.18
$[C_5H_5NO]_2SnCl_4$	0.42		9.29
$[Cl_2SeO]_2SnCl_4$	0.37		9.64
$(C_2H_5O)_2P(O)CH_2CO_2C_2H_5SnCl_4$	0.37	0.69	9.64
$CH_3OCCH = CH.CO\ CH_3.SnCl_4$	0.38	0.91	9.57
$[C_6H_5.CH = CH.COCH_3]_2SnCl_4$	0.38	1.09	9.57
$(CH_3OH)_2.SnCl_4$	0.43	0.70	9.22
$(C_2H_5OH)_2SnCl_4$	0.33	0.70	9.92
$(C_3H_7OH)_2SnCl_4$	0.43	0.50	9.22
$[(CH_3)_2CHOH]_2SnCl_4$	0.35	0.70	9.78
$[\{(CH_3)_2N\}_2CO]_2SnCl_4$	0.35	0.75	9.78
$[(C_2H_5)_2O]_2SnCl_4$	0.45	1.10	9.09
$[(CH_2)_4O]_2SnCl_4$	0.43	1.14	9.22
$[(CH_3)_2CO]_2SnCl_4$	0.40	small	9.43
$[(CH_2)_5CO]_2SnCl_4$	0.35	.0	9.78
$C_2H_5O.(C_2H_5OH)_2SnCl_3$	0.33	small	9.40
$\{[(CH_3)_2N]_3PO\}_2SnCl_4$	0.31	0.70	10.06
$[(CH_3)_2N]_3PO\ _2SnBr_4$	0.56	0.76	9.71
$[(C_6H_5)_3PO]_2\ SnBr_4$	0.63	0.61	9.22
$[(CH_3)_2SO]_2SnBr_4$	0.66	–	9.02
$(C_6H_5O.CO_2)_2SnCl_2$	0.23	0	9.45
$(C_6H_5O.CO_2)_2SnBr_2$	0.28	0	9.63
$[(C_6H_5O.CO_2]_2SnI_2$	0.41	0	9.59
$(C_6H_5OHCO_2H)_2SnCl_4$	0.43	1.10	9.23
$(C_6H_5OH.CO_2H)_2SnBr_4$	0.73	1.22	8.53

Donor Atom Nitrogen

$(C_{10}H_8N_2)SnCl_4$	0.47	0.00	8.95
$(C_{10}H_8N_2)SnBr_4$	0.73	0.00	8.53
$(C_5H_5N)_2\ SnCl_4$	0.51	0.00	8.67
$(C_5H_5N)_2SnBr_4$	0.74	0.00	8.46
$(CH_3)_2N(CH_2)_2N(CH_3)_2SnCl_4$	0.61	0.00	7.97
SnF_2pc	0.03	0.70	9.12
$SnCl_2pc$	0.28	0.99	9.27
$SnBr_2pc$	0.34	1.09	9.40
SnI_2pc	0.45	0.99	9.40
$SnCl_2TMeoxpp$	0.24		9.40
$Sn(OH)_2TMeoxpp$	0.19		9.40(PP)
			9.60(OH)

Donor Atom S

$[((CH_3)_2N)_2CS]_2SnCl_4$	0.70	0.97	7.34
$[((CH_3)_2N)_2CS]_2SnBr_4$	0.94	0.92	7.04
$[CH_3S(CH_2)_2S\ CH_3]SnCl_4$	0.70	0.87	7.34
$[CH_3S(CH_2)_2SCH_3]SnBr_4$	0.97	0.89	6.84
$[C_4H_8S]_2\ SnBr_4$	0.99	0.84	6.71
$[(C_5H_{10}N)_3PS]_2.SnCl_4$	0.49	1.05	8.85
$[(C_4H_8N)_3PS]_2\ SnCl_4$	0.43	1.06	9.22
$[(C_4H_8N)_3PS]_2\ SnBr_4$	0.61	1.20	9.30,

Donor Atom P

$[(n-C_4H_9)_3P]_2SnCl_4$	0.87	1.02	6.16
$[(C_2H_5)_3P]_2SnCl_4$	0.97	0.95	6.16
$[(C_3H_7)_3P]_2SnCl_4$	0.89	0.95	6.00
$[(CH_3)_2C_6H_5P]_2SnCl_4$	0.85	0.97	6.25
$[CH_3(C_6H_5)_2P]_2SnCl_4$	0.81	0.58	6.60
$[CH_3O(C_6H_5)_2P]_2SnCl_4$	0.81	0.45	6.60
$[(C_6H_5)_3P]_2SnCl_4$	0.75	0	6.90
$[(C_6H_5)_2PC \equiv CCH_3]_2SnCl_4$	0.77	0	6.80

Donor Atoms Nitrogen and Oxygen

Compound	δ	ΔE_q	$\bar{X}_{AV}$
$SnCl_2ox_2$	0.32	0	9.14
$SnBr_2ox_2$	0.44	0	9.07
SnI_2ox_2	0.61	0	8.90
$SnCl_3ox\ oxH$	0.37	0	9.19
$SnF_3ox_2H_2F$	-0.08	0	9.19
$SnCl_4(oxH)_2$	0.45	0	9.09
$SnBr_4(oxH)_2$	0.65	0	9.09
$SnI_4(oxH)_2$	0.91	0	8.95
$SnCl_3oxPyHCl$	0.40	0	9.10(ox)
			8.66(Py)
$SnCl_2pic_2$	0.21		9.17
$SnBr_2pic_2$	0.34		9.05
SnI_2pic_2	0.45		9.10
$Sn(OH)_2pic$	0.09		9.10(pic)
			9.70(OH)

Data from ref. 9, 13-18
pc = phthalocyanine, picH = picolinic acid, py = pyridine
OxH = 8-hydroxy quinoline
TMeoxpp = Tetra [methoxy phenyl] porphorin

$$F > O > N > Cl > Br > C \underset{\sim}{} S > P > Sn$$

and O $R_3PO \simeq R_2SO > O > \overline{OH} > -OMe > Cl_3PO*$

N $N > N\langle\!\langle\bigcirc\!\rangle\!\rangle > N\langle\!\langle\bigcirc\!\rangle\!\rangle > N R_2$

$P = S > C = S > S > R_2S$

$P > \phi_3P > \phi_2RP > R_3P$.

SnX$_4$, LSnX$_4$ and SnX$_5^-$ compounds. The examination of δ values on SnX$_4$ compounds leads to a similar relationship of δ with the electronegativity of ligand/anion. Again, by comparison with the theoretical calculations and the experimental results on SnX$_4$ compounds we derive a simple correlation of δ with $\bar{X}$ which is applicable over $\delta = 1.0 - 1.4$ mm.sec^{-1}

$$\delta = -0.106 \sum_1^4 \bar{X}_i + 4.35 \qquad\qquad [3]$$

Equation [2] and [3] have the same linear dependence with $\frac{1}{n} \sum_n \bar{X}_i$ but the intercepts differ by 0.22 mm.sec^{-1}. Clausen and Good find greater dependence of $\frac{1}{n} \sum X_i$ than Parish and Platt. However, the difference is minor and perhaps dependent on the choice of δ values for compounds for which the coordination number is assumed to be four with tetrahedral geometry through the involvement of sp^3 orbitals. In Table 2, we include a few compounds giving $\bar{X}$ for the ligands. The agreement between the values of $\bar{X}$ obtained for L$_n$SnX$_{6-n}$ compounds and for L$_n$SnX$_{4-n}$ compounds is surprisingly good and is consistent with the earlier concept that δ is determined by the electronegativity of the nearest neighbours of the Mossbauer atom in a molecule. The isomer shifts of $\bar{X}_x$ calculated for N,P,S,Sn from $(R_3Sn)_3X$ agrees well with the accepted values of electronegativities (19). The differences in the calculated values of $\bar{X}$ are found for those ligands for which $|\bar{X}_A - \bar{X}_B| > 1.5$. This is to be expected because contributions due to population of p orbitals has not been taken into account. It is interesting to note that the isomer shift is lower by 0.22 mm.sec^{-1} when coordination number increases from 4 to 6. It is known that the Sn-X bond lengths increase in going from SnX$_4$ to SnX$_6^{2-}$ which is attributed to greater ionicity of the σ bonds in SnX$_6^{2-}$ compounds. Some subsidiary π interaction has been indicated in SnX$_4$ compounds by Goldanskii et al (6) which leads to shortening

* This appears to be contrary to the expected order of electronegativities of R$_3$PO and Cl$_3$PO (20,21). However, the result expresses the idea that Cl$_3$PO interacts more weakly with Sn(IV) than R$_3$PO.

Table 2

Electronegativity of R and X in $R_n SnX_{4-n}$
Compounds

Compounds	δ mm.sec^{-1}	ΔE_q	$\bar{X}_R$	X_X
SnS$_2$	1.20			7.40
SnCl$_4$	0.80			8.35
SnBr$_4$	1.13			7.51
SnI$_4$ ·	1.45			6.81
(CH$_3$)$_4$Sn	1.29		7.20	
(n-C$_4$H$_9$)$_4$Sn	1.35		7.07	
(C$_6$H$_{11}$)$_4$Sn	1.52		6.67	
(C$_6$H$_5$)$_4$Sn	1.21		7.40	
(C$_6$F$_5$)$_4$Sn	1.10		7.65	
[(CH$_3$)$_3$]Sn$_2$	1.46		7.20	5.7(1.96)*
Sn(Gray)	2.10			5.3
(CH$_3$)$_2$SnO	0.92	2.0	7.20	9.0
(C$_6$H$_5$)$_2$SnO	0.88	~2.0	7.40	8.95
[(C$_6$H$_5$)$_3$]SnO	1.08	2.15	7.40	9.0
(CH$_3$)$_2$Sn(OCH$_3$)$_2$	0.99	2.31	7.20	8.63
(C$_6$H$_5$)$_3$SnO.CON(CH$_3$)$_2$	.04	1.58	7.40	9.05
[(CH$_3$)$_3$Sn]$_3$N	1.10	small	7.20	9.0 (3.04)
(CH$_3$)$_3$SnNR$_2$	~1.14	0-1.1	7.20	8.67
(C$_6$H$_5$)$_3$SnNR$_2$	~1.10	0	7.40	8.49
[(CH$_3$)$_3$Sn]$_3$P	1.33	0	7.20	6.90(2.32)
[(CH$_3$)$_3$Sn]$_3$As	1.35	0	7.20	6.70(2.28)
RSn S$_{1.5}$	1.38	1.45	7.20	6.90
[(C$_6$H$_5$)$_3$Sn]$_2$S	1.22	1.17	7.40	7.35
(C$_6$H$_5$)$_3$Sn.Re(CO)$_5$	1.45	0	7.40	6.13
C$_6$H$_5$Sn$_5$[Re(CO)$_5$]$_3$	1.75	0	7.40	5.72

* Converted to Pauling's scale.
  Data from ref. 1 and 3.

of Sn-X bonds. In all, δ is known to be dependent on three effects namely, a) the ionicity of σ bonds, b) the population of p and d orbitals in the valence states and c) π interaction. Effect a) will reduce the isomer shift by increasing the effective nuclear charge whereas effect b) and c) will increase or decrease the isomer shift through d $\pi \leftarrow$ pπ or d $\pi \rightarrow$ dπ bonding by screening of 5 s electron. Again it has been shown by Ruby et al that the shielding effect due to the population of p orbitals is small ($\sim$0.1 mm.sec^{-1}) when relativistic corrections are applied and for the population of d orbitals it is still smaller, 5 d orbitals in tin atoms being diffuse. The decrease in isomer shift for 6 coordinated compounds can be attributed to the increase in the ionicity of Sn-X or Sn-L bonds. In other words, an increase in coordination number reflects in lower s electron density at the nucleus. From the data on $Sn(NO_3)_4$, $Sn(ox)_4$ which are known to be 8 coordinated compounds, the isomer shift as obtained from eqn. (3) is higher by ~ 0.43 mm.sec^{-1} compared to the experimental value (22). A decrease in coordination number from 8 to 6 for $Sn(NO_3)_4$ and $Sn(NO_3)_4$ Py_2 is indicated by the corresponding increase in δ (0.20 mm.sec^{-1}) [23]. A similar increase in δ is observed for $Sn(NO_3)_4$ bipy. From the experimental data on 5 and 7 coordinated compounds, the following correlations are suggested

$$\delta = \frac{0.430}{n} \cdot \sum_{i}^{n} \bar{X}_i + C \qquad [4]$$

where C= 4.25 when n = 5, C = 4.02 when n = 7, C = 3.94 when n = 8, n = coordination number. The calculated values of δ along with δ_{exp} are presented in Table 3. The values

Table 3

SnXY compounds

Coordination No.8

Compound	$\bar{X}$	δ_{exp}	δ_{cal}
$Sn(ox)_4$	9.10	0.04	0.05
$Sn(NO_3)_4$	9.28	0.00	-0.01
$Snpc_2$*	9.30	0.11	-0.05

*Probably CN = 6

Coordination No.7

$SnOx_3Cl$	9.10	0.11	0.12
$Sn\ pic_3Cl$	9.10	0.18	0.18
$Sn\ pic_3Br$	9.10	0.15	0.15
$Sn(Dipic)_2H_2O$	8.66(py) 9.0(H_2O)	0.19	0.20

Data from ref. 9 and 18
Dipic = Dipicolinic acid.

of δ for SnO_2 and SnF_4 are consistent with their known
structures as 6 coordinated compounds. Application of
equations [2] and [4] thus allows a prediction of the coordination number of a metal in a complex to be made from
the value of the Sn-119 isomer shift of the compound. The
isomer shift (0.27 mm.sec^{-1}) for $NOSnCl_4NO_3$ indicates that
the NO_3^- ion is certainly bidentate. Similarly, the compounds
SnX_2 $\{[Ph_2P(O)]_2N\}_2$ have a coordination number of 8.
$[(C_6H_5)_3P]SnCl_4$ is predicted to be a dimer in the solid state.

Organotin (IV) Compounds. The Mossbauer parameters of
organotin (IV) compounds often do not conform to any of the
correlations presented in eqn. [2]-[4] . The isomer
shift increases when R = alkyl or aryl is replaced by a
more electronegative group in R_4Sn compounds (Table 5).
Parish and Platt attribute this change to a deshielding and
contraction of 5 s electrons resulting from a greater involvement of p electrons in bonding to a more electronegative
group rather than rehybridization at the tin atom (12).
However, there are many exceptions where the above concept
fails to account for the observed isomer shifts, notably in
$Ph_3Snox(\delta = 1.07)$, $Me_3SnOH(\delta = 1.07)$, Me_3SnNR ($\delta = 1.16$),
$Ph_3SnOH(\delta = 1.18)$ where X is more electronegative than Cl.
Herber and coworkers (8, 24, 25) have found a correlation
of isomer shift with electronegativity of X in $(CH_3)_3SnX$
compounds. Unfortunately, the correlation is not applicable
to all R_3SnX compound viz. $(CH_3)_3SnNO_3$. (23).

If Ruby et al's calculations on the population of s and
p orbitals in sp^2 and sp hybrid orbitals at the tin atom are
followed, the correlation of isomer shift with electronegativity can be extended to situations where 5 s electrons preferentially occupy either sp^2 or sp hybrid orbitals and the
remaining p orbitals, as may be the case in sp^2 and sp
hybridisation, form Sn-X bonds. The involvement of p
orbitals alone affects the isomer shift in a similar manner
to that outlined earlier (0.12 mm.sec^{-1} per bond)

$$\delta = -0.153 \sum_1^3 X_i - 0.04 \sum_j (X_j - X_{Sn}) + 4.80-(n-4)K \quad [5]$$

for sp^2 hybrids

$$\delta = -0.25 \sum_1^2 X_i - 0.04 \sum_j (X_j - X_{Sn}) + 5.20-(n-4)K \quad [6]$$

for sp hybrids

X_i = electronegativity of ligands forming bonds through
hybrid orbitals of the tin atom.

$\bar{X}_i$ = electronegativity of ligands forming bond through p orbitals of the tin atom

$\bar{X}_{Sn}$ = electronegativity of Sn   and

n = coordination number.

While it can be argued that pure sp^2 or sp hybrid orbitals and pure p orbitals do not exist in tin compounds, x-ray crystallographic data as presented in Table 4, strongly

Table 4

Structural parameters of $L_nR_mSnX_{4-m}$

	<C-Sn-C	<C-Sn-X	X-Sn-X	Geometry
$(CH_3)_2SnCl_2$[a]	123.3	109.0°	93.0°	dis.tet.
$(CH_3)_2Sn(CN)_2$[b]	148.7		85.3	" "
$(CH_3)_2Sn(NCS)_2$[c]	145.9	90.7	84.1	" "
$BuSnS_{1.5}$[d]		109	109	tet.
$(CH_3)_3SnF$[e]	~120°	~90°	<180	trigbip.
$(CH_3)_3SnClpy$[f]	~120			trigbip.
$(CH_3)_3SnCN$[g]	~120	~90°	<180	trigbip.
$(CH_3)_3SnNCS$[h]	120°	~90	<180	trigbip.
$(C_6H_5CH_2)_3SnO.COCH_3$[i]	123.8	90	168.6	trigbip.
	108.1			
	125.5			
$(CH_3)_2SnCl_3^-$[f]	~120			trigbipy
$(CH_3)_2SnF_2$[j]	180	90	90 & 180	Oct.
$(CH_3)_2SnCl_2.2(CH_3)_2SO$[k]	180	90	87 & 180	Oct.
$(CH_3)_2SnCl_2 2C_5H_5NO$[l]	180	90	90 & 180	Oct.
$(CH_3)_2Snox_2$[m]	110			

a ref 26 b ref 27 c ref 28 d ref 29 e ref 30 f ref 31 g ref 32 g ref 33 h ref 34 i ref 35 j ref 36 k ref 37 l ref 38 m ref 39

suggest preferential involvement of s electrons in hybrid orbitals. Herber, Drago and Hill (24) have suggested that methyl moieties are bonded through sp^2 hybrid orbitals of the Sn atom in $(CH_3)_3SnX$ compounds and X is bonded through a p orbital of Sn. According to Whitehead and Jaffe (40) the involvement of s electrons in BCl_3, PCl_3 $AsCl_3$ depends on the bond angles (33% for 120°, 18% for 103°, 14.8% for 100°). In $(CH_3)_2Sn(CNS)_2$, $(CH_3)_2SnCl_2$ and $(CH_3)_2Sn(CN)_2$ the X-Sn-X angle (<95°) suggests minimal involvement of s orbitals in Sn-X bonds. In R_2SnX_2 compounds δ can range from 1.4-2.0

mm.sec^{-1} depending on the $\bar{X}_R$ value and the coordination
number. The isomer shift in R_3SnX compounds in which the
compounds are known to have trigonal bipyramidal geometry in
the solid state (30-35) with R in the trigonal plane can be
calculated according to eqn. [5]. The δ values for R_3SnX
can range from 1 mm.sec^{-1} to 1.7 mm.sec^{-1} depending on the
hybrid orbitals employed in the bonding and the coordination
number.

Table 5

R_3Sn X Compounds

Compounds	δ mm.sec^{-1}	ΔE_q mm.sec^{-1}	$\bar{X}_R$	$\bar{X}_x$
$(CH_3)_3SnF$	1.25(1.31)	3.80	7.2	10.41
$(CH_3)_3SnCl$	1.41(1.39)	3.32	7.2	8.29
$(CH_3)_3SnBr$	1.4 (1.42)	3.30	7.2	7.59
$(CH_3)_3SnI$	1.48(1.46)	3.05	7.2	6.75
$(CH_3)_3SnCN$	1.36(1.36)	3.22	7.2	9.0
$(CH_3)_3SnNO_3$	1.44(1.35)	4.14	7.2	9.28
$(C_6H_5)_3SnF$	1.25(1.21)	3.53	7.4	10.41
$(C_6H_5)_3SnCl$	1.31(1.29)	2.56	7.4	8.29
$(C_6H_5)_3SnBr$	1.37(1.32)	2.48	7.4	7.59
$(C_6H_5)_3SnI$	1.20(1.36)	2.25	7.4	6.75
$(C_6H_{11})_3SnF$	1.56(1.55)	3.96	6.67	10.41
$(C_6H_{11})_3SnCl$	1.64(1.64)	3.49	6.67	8.29
$(C_6H_{11})_3SnBr$	1.63(1.66)	2.90	6.67	7.59
$(C_6H_{11})_3SnAc$	1.57(1.61)	3.27	6.67	9.0

The experimental and the calculated values for typical compounds are presented in Table 5 for assumed sp^2 hybridization
scheme in R_3SnX compounds. In our calculations, we have
assumed ideal geometries but in some compounds such is not
the case. There are minor distortions perhaps due to steric
effects and additional terms may be required to calculate
the isomer shift with respect to the disposition of ligands
around the tin atom. Moreover, a large scatter in the reported isomer shift values on compounds does not permit a
better correlation. However, the basis of the model for
predicting geometry seems to relate to one consistent parameter i.e. the molecular electronegativity. An interesting
case is that of $(CH_3)Sn(NO_3)_3$ which merits some discussion.
The crystallographic data indicates that it is a 7 coordinated

compound with C–Sn–O angles of $\sim90°$ and $152°$ [41]. 5 oxygen atoms are in a pentagonal plane perpendicular to the Sn–C axis. If we assume the involvement of sp hybrid orbitals of tin for the formation of Sn–C and Sn–O bonds and p^2 and perhaps d orbitals to form the rest of Sn–O bonds, use of eqn. [6] taking into account that the coordination number is 7, gives δ_{cal} (0.9 mm.sec^{-1}) which is in good agreement with δ_{exp} = .94 mm.sec^{-1}. It can be predicted from eqn. [3]–[6] that the isomer shift in $RSnX_3$ will show a greater dependency with respect to the electronegativity of X than in the corresponding R_3SnX and R_2SnX_2 compounds when R and X form bonds through any set of orbitals (Table 6).

Table 6

$RSnX_3$ Compounds

Compound	δ mm.sec^{-1}	ΔE_q mm.sec^{-1}	$\bar{X}_x$
n-$C_4H_9SnCl_3$	1.38	1.86	8.29
n-$C_4H_9Sn(OH)_2Cl$	0.74	2.02	9.50
n-$C_4H_9Sn(ox)_2Cl$	0.78	1.67	9.10
CH_3SnBr_3	1.41	1.91	7.59
$CH_3Sn(NO_3)_3$	0.94	1.90	9.28

$(CH_3)_3SnNO_3$ and its adducts are known to have a trigonal bipyramidal geometry with oxygen atoms in the axial position (42,43). δ = 1.44 $\pm$ 0.10 mm.sec^{-1} are in agreement if we assume that sp^2 hybrid orbitals are involved in the formation of Sn–C bonds. It may be mentioned that the molecular electronegativity of NO_3^- ion cannot be predicted from the correlation proposed by Herber and Leung (25).

$R_nSnox_mX_{4-n-m}$ compounds present an interesting series of compounds for which Mossbauer parameters have been obtained. It has been suggested that cis and trans isomers can be predicted from the observed values of δ (44). However, trans isomers have not been prepared. The crystal structure of $(CH_3)_2Sn(ox)_2$ shows that < C–Sn–C is $110°$ and that oxygen and nitrogen atoms in the 8 hydroxy quinolate moiety lie in a tetrahedral plane. Schlemper (39) has discussed the bonding in terms of using sp^3 hybrid orbitals at the tin atom with the assumption that one of the sp^3 orbitals is used in bonding with both N and O atoms. The Mossbauer parameters support the above view since δ_{exp} agrees with the calculated one following eqn. [3] (Table 7). It may be argued that eqn. [3] should be applicable for

Table 7

$R_n Sn(ox)_m X_{4-n-m}$ Compounds

Compound	δ mm.sec^{-1}	ΔE_q mm.sec^{-1}	Hybrid orbitals of Sn
$C_6H_5Sn(ox)_2Cl$	0.66	1.48	sp^3
$(C_6H_5)_2Sn(ox)_2$	0.78	1.64	sp^3
$(C_6H_5)_3Snox$	1.07	1.75	sp^3
$n-C_4H_9\ Sn(ox)_3$	0.68	1.70	sp^3
$(n-C_4H_9)_2Sn(ox)_2$	0.93	2.05	sp^3
$(n-C_4H_9)_2Sn(ox)Cl$	1.40	3.21	sp^2,p
$(CH_3)_2Sn(ox)_2$	0.84	1.9	sp^3
$(CH_3)_2Sn(ox)Cl$	1.26	3.12	sp^2,p
$(C_2H_5)_2Sn(ox)_2$	0.87	2.03	sp^3
$(C_2H_5)_2Sn(ox)Cl$	1.34	3.13	sp^2,p
$(C_2H_5)_2Sn(ox)Br$	1.39	3.08	sp^2,p
$(C_2H_5)_2Sn(ox)I$	1.43	2.85	sp^2,p

calculating $\bar{X}_{ox}$ in $SnCl_2(ox)_2$. $SnCl_2(ox)_2$ is isomorphous
with $TiCl_2(ox)_2$. The bond angle Cl-Ti-Cl is 96° suggesting
that the s orbital involvement is minimal if sp^3 hybrid
orbitals are employed (45). The disposition of ligands
around the tin atom can be indicated on the basis of the
isomer shift measurements. $[R_nSnO(OH)]_n$ compounds show
interesting results in terms of isomer shifts. $[CH_3\ SnO(OH)]_n$
has an isomer shift of 0.4 mm.sec^{-1} indicating O and OH are
bonded through sp^2 hybrid orbitals of tin atoms, whereas,
for $[C_2H_5SnO(OH)]_n$ the isomer shift is 0.7 mm.sec^{-1} indicating O or O and OH and C_2H_5 are bonded through sp^2 hybrids
of tin atoms. Bent's rule (46), that s character is
'preferentially' concentrated in bonds with more electro-
positive ligands, is not followed. Similarly, Mossbauer
parameters of the compounds $SnOX_2$ (X = halogen) indicate
that Sn-O bonds are formed through sp hybrid orbitals whereas
bonds with Cl or Br or I are formed with p orbitals. Clearly,
x-ray crystallographic studies on these compounds are de-
sirable. If Mossbauer spectroscopy is to become a standard
technique for chemists, one should be able to deduce information like bond angles, bond lengths from Mossbauer
parameters. It may be emphasized that these correlations
be treated with caution since steric effects are not taken
into account for calculating δ.

Quadrupole splitting. Two opposing models in terms of bonding have been proposed for the interpretation of the observed ΔE_q in Sn(IV) compounds. The model of Greenwood and Ruddick (47) ascribes ΔE_q as due to subsidiary π interactions. In recent publications, the Greenwood and Ruddick rule has been found invalid and σ bond effects are considered more important in causing asymmetry in the electric field gradient (efg) at the nucleus. Fitzsimmons et al and Parish and Platt (44,50,51) have followed the point charge model (48,49) for the interpretation of the observed ΔE_q in organotin halides. The limitations of the point charge model for predicting ΔE_q for organotin compounds have been pointed out by Platt (53).

It is known that the efg is caused mainly by the imbalance in the populations of orbitals. Whitehead and Jaffe (40) by following the theory of Townes and Dailey (52) have interpreted nuclear quadrupole resonance frequencies in terms of orbital population of s and p electrons in halides. They have related the ratio, ρ, of the resonance frequency in a compound to that of the atom by the equation

$$\rho = (1 - s + d)(1 \pm i) - \pi \tag{7}$$

where s and d represent s and d hybridization in an orbital and π allows for π overlap. i is the ionic character and according to Gordy is equal to $\dfrac{|\bar{X}_A - \bar{X}_B|}{2}$.

For tin-119^m with a spin of 3/2, ΔE_q is given by the expression

$$\Delta E_q = \tfrac{1}{2} e^2 q \, Q \, (1 + \eta^2/3)^{\tfrac{1}{2}} \tag{8}$$

where eq = electric field gradient in the z direction

Q = nuclear quadrupole moment

$$\eta = \frac{|V_{XX}| - |V_{YY}|}{|V_{ZZ}|} = \text{asymmetry parameter}$$

In trans L_2SnX_4 compounds having octahedral geometry, the asymmetry parameter vanishes. ΔE_q can then be related to the electron populations of tin ion valence states accor-

ding to eqn. [2]. The value of s can be obtained from the
isomer shift (eqn. [2]-[5]) but there is very little agreement on the evaluation of i. However, it can be predicted
that the compounds having trans geometries in which
$|\bar{X}_L - \bar{X}_x| > 1$ will give observable ΔE_q. Non-observance of ΔE_q in
these compounds indicates additional interactions which may
or may not be due to steric or π-effects. In L_2SnCl_4 compounds, ΔE_q is observed for ligands having oxygen or phos-
phorus atoms (13,15) where as ligands with nitrogen as donor
the atom ΔE_q is negligible.

Recently Williams and Kocher (53) and Devooght et al
(54) have derived expressions for ΔE_q for organotin compounds
based on bond polarities and sp^3 hybridization. The agreement between the observed and the calculated values could
be obtained only on the assumption of d orbital population.
There is very little direct experimental evidence in Sn(IV)
compounds for back bonding. Attempts are being made to
derive expressions for ΔE_q on the assumption that different
sets of orbitals are involved in bonding. Our preliminary
results on the derivation of ΔE_q for R_3SnX compounds show
good agreement with the experimental values but the model
calculations need to be extended to other compounds namely
$RSnX_3$ and R_2SnX_2.

Eu(III) complexes. The isomer shift for Eu(III) chlo-
ride, acetate, perchlorate and sulfate and their adducts
with ligands of N and O donor atoms were determined in our
laboratory. The isomer shift with respect to Eu_2O_3 as a
source was found to be in the range of 0-1 mm.sec^{-1} for all
the compounds with the coordination number changing from 6
to 10. The adducts with N atom donors gave a lower value of
δ than for adducts with O as the donor atom (56).

It is known that the isomer shift of Eu shows a large
variation in δ value when its valence state changes from Eu^{3+}
to Eu^{2+}. Furthermore, the isomer shift shows a large variation in Eu(II) compounds viz. in $EuSO_4 \delta = 15.4$ mm.sec^{-1} and
in EuO $\sim$11 mm.sec^{-1}. It would appear that Eu(II) shows a
large change in s electron density at the nucleus with changes
in molecular electronic structure. On the other hand, Eu(III)
shows very little change in δ and thus its interaction with
ligands/anion seems minimal in terms of electronic involvement. A larger value of δ for compounds having ligands with
oxygen as the donor atom seems to indicate a greater electronic involvement with 'hard' bases.

It has been shown by Greenwood et al, (57) that the
shape of the resonance absorption spectrum will alter from
a Laurentzian one if there is non-cubic symmetry at the Eu
nucleus. The examination of spectra indicated very little
deviation from the Laurentzian shape and as such it appears
that the ligands or anions are not able to alter efg at the
nucleus. Our results on all the compounds have provided
little information which can be of any significance to a
structural chemist.

In conclusion, the isomer shift of Sn(IV) compounds has
been interpreted in terms of the molecular electronegativity
of ligands or anions forming bonds with tin. The x-ray
crystallographic data is lacking to test the prediction of
geometry and bonding parameters from the isomer shift data.
The quadrupole splitting is helpful in substantiating the
predictions from the isomer shifts.

ACKNOWLEDGEMENT

The authors are grateful to the National Research
Council of Canada for providing financial support for this
research.

REFERENCES

1.   P.J. Smith, Organometal.Chem. Rev., A, 5, 373 (1970).

2.   V.I. Goldanskii, V.V. Khrapov and R.A. Stukan, Organo-
      metal. Chem. Rev., A, 4, 225 (1969).

3.   V.I. Goldanskii and R.H. Herber, 'Mossbauer Effect and
      Its Application in Chemistry'. Consultants Bureau
      Enterprises, Inc., New York, (1964).

4.   J.J. Spijkerman, The Mossbauer Effect and Its Application
      in Chemistry Ed. R.F. Gould A.C.S. Pub. (1967).

5.   J.K. Lees and P.A. Flinn, Phys. Letters 19, 186 (1965);
      J.K. Lees and P.A. Flinn, J. Chem. Phys., 48, 882 (1968);
      R.H. Herber, H.A. Stockler and W.T. Reichle, J. Chem.
      Phys., 42, 2447 (1965).

6.   V.I. Goldanskii, E.F. Makarov and R.N. Stukan, J. Chem.
      Phys., 47, 4048 (1967); V.I. Goldanskii, E.F. Makarov,
      R.A. Stukan, T.N. Sumarokova, V.A. Trukhtanov and
      V.V. Khrapov, Doklady Akad. Nauk. SSSR, 156, 474 (1964).

7.   S.L. Ruby, G.M. Kalvius, G.B. Beard and R.E. Snyder, Phys. Rev., 159, 239 (1967).

8.   H.S. Cheng and R.H. Herber, Inorg. Chem., 9, 1686, (1970).

9.   K.M. Ali, D. Cunningham, M.J. Frazer, J.D. Donaldson and B.J. Senior, J. Chem. Soc., (A) 2836, 1969.

10. C.A. Clausen and M.L. Good. Inorg. Chem., 9, 817 (1970).

11. A.G. Davies, L. Smith and P.J. Smith, J. Organometal. Chem., 23, 135 (1970).

12. R.V. Parish and R.H. Platt, Inorganica Chimica Acta, 4, 589 (1970).

13. A.J. Carty, T. Hinsperger, L. Mihichuk and H.D. Sharma, Inorg. Chem., 9, 2573 (1970).

14. J. Phillip, M.A. Mullins and C. Curran, Inorg. Chem., 7, 1895 (1968).

15. P.A. Yeats, J.R. Sams and F. Aubke, Inorg. Chem., 9, 740, 1970.

16. S. Ichiba, M. Mishima, H. Sakai and H. Negita, Bull. Chem. Soc. Japan, 41, 49 (1968).

17. D.V. Naik and C. Curran, Inorg. Chem., 10, 1017 (1971).

18. M. O'Rourke and C. Curran, J. Am. Chem. Soc., 92, 1501 (1970).

19. L.E. Allred, J. Inorg. Nucl. Chem., 17, 43 (1961), ibid, 17 215 (1961).

20. J. Hinze and H.H. Jaffe, J. Phys. Chem., 67, 1501 (1963).

21. J.E. Huheey, J. Phys. Chem., 70, 2086 (1966).

22. G.D. Garner, D. Sutton and S.C. Wallwork, J. Chem. Soc., (A), 1949 (1967).

23. A.J. Carty and H.D. Sharma, Unpublished results.

24. J.C. Hill, R.S. Drago and R.H. Herber, J. Am. Chem. Soc.,
    91, 1644 (1969).

25. B. Gassenheimer and R.H. Herberg, Inorg. Chem., 8, 1120
    (1969). K.L.Leung and R.H.Herber,Inorg.Chem.,10,1020(1971).

26. A.G. Davies, H.J. Milledge, D.C.Puxley and P.J. Smith,
    J. Chem.Soc., (A) 2862 (1970).

27. J. Konnert and D. Britton, Unpublished work.

28. Y.M. Chow, Inorg. Chem., 9, 794 (1970). J.B. Hall as
    quoted in the above.

29. C. Dorfelt, A. Janeck, D. Kobelt, E.F. Paulus and
    H. Scherer, J. Organometal, Chem., 14, 22 (1968).

30. H.C. Clark, R.J. O'Brien and J. Trotter, J. Chem. Soc.,
    2332 (1964).

31. F.W.B. Einstein and B.R. Penfold, J. Chem. Soc., (A),
    3019 (1968).

32. R. Hulme, J. Chem. Soc., 1524 (1963).

33. E.O. Schlemper and D. Britton, Inorg. Chem., 5, 507
    (1966).

34. R.A. Forder and G.M. Sheldrick, J. Organometal. Chem.,
    22, 611(1970), 21 115(1970).

35. N.W.Alcock and R.E.Timm. J. Chem. Soc., A, 1873 (1968),
    J. Chem. Soc., A, 1876 (1968).

36. E.O. Schlemper and W.C. Hamilton, Inorg. Chem., 5, 995
    (1966).

37. N.W. Isaacs, C.H.L. Konnard and W. Kitching, Chem.
    Comm., 820 (1968).

38. E.A.Blom, B.R. Penfold and W.T. Robinson, J. Chem.
    Soc., (A), 913 (1969).

39. E.O. Schlempler, Inorg. Chem., 6, 2012 (1967).

40. M.A. Whitehead and H.H. Jaffe, Theor.Chim.Acta,1,209(1963).

41. C.S. Nyberg, J.T.Szymanski, G. Brownlee, D. Potts and
     A. Walker, Chem. Comm. In press.

42. P. Au, M.Sc. Thesis, University of Western Ontario,
     London, Ontario.

43. D. Potts, Ph.D. Thesis, University of Toronto, Toronto,
     Ontario.

44. B.W. Fitzsimmons, N.J. Sealey and A.W. Smith, J. Chem.
     Soc., (A), 143 (1969).

45. B.F. Studd and A.G. Swallow, J. Chem. Soc., (A), 1961
     (1968).

46. H.A. Bent, Chem. Rev; 61, 275 (1961).

47. N.R. Ruddick and N.N. Greenwood, J. Chem. Soc., (A)
     1679 (1967).

48. R.W. Grant, Mossbauer Effect Methodology, 2, 23 (1966).

49. R.L. Collins and J.C. Travis, Mossbauer Effect Metho-
     dolgy, 3, 123 (1967).

50. R.V. Parish and R.H. Platt, J. Chem. Soc., (A), 2145
     (1969).

51. R.V. Parish and R.H. Platt, Inorganica Chimica Acta.,
     4, 65 (1970).

52. B.P. Dailey and C.H. Townes, J. Chem. Phys., 23, 118,
     (1955).

53. D.E. Williams and C.W. Kocher, J. Chem. Phys., 52, 1480
     (1970).

54. J. Devooght, M. Gielen and S. Leieune, J. Organometal.
     Chem., 21, 333 (1970).

55. R.H. Platt, J. Organometal, Chem., 24, C23 (1970).

56. A.P. Beckler, M.Sc. Thesis, University of Waterloo,
     Waterloo, Ontario.

57. B.A. Goodman, N.N. Greenwood and G.E. Turner, Chem.
     Phys. Letters, 5, 181 (1970).

ELECTRIC-FIELD GRADIENT (EFG) CALCULATIONS IN "IONIC" CRYSTALS

J. O. Artman

Carnegie Institute of Technology and Mellon Institute of Science, Carnegie-Mellon University, Pittsburgh, Pennsylvania 15213

I. PRELIMINARIES[1]

The electrostatic potential energy of a binary point charge system is

$$W = \frac{q_1 q_2}{|r_1 - r_2|} . \tag{1}$$

If we generalize Eq. (1) to charge distributions we have

$$W = \int_{\tau_2} \int_{\tau_1} \frac{\rho_1(r_1)\rho_2(r_2)d\tau_1 d\tau_2}{|r_1 - r_2|} . \tag{2}$$

In this paper we are concerned with a relatively small charged body (an atomic nucleus) in the field of its charged ambient. The nucleus of interest to us (as well as its companions) is distributed in space in accordance with the geometrical specifications of a crystal "structure". For the time being we adopt the "ionic" model, by which we mean simply that we regard each nucleus and its associated electrons as a separate system of charges distinct from the other such systems in the crystal. We consider the centroid of the nucleus of interest as a fiducial point; developing the electrostatic interaction with respect to this reference will necessarily involve the use of nuclear "multipoles". Therefore, we relate τ_1 and τ_2 in Eq. (2) to the nuclear and "external" charge volumes τ_n and τ_e respectively and proceed. (Although s-electrons penetrate the nucleus, they do not contribute to quadrupole effects; such "contact" terms will be ignored here.) The region τ_e may be effectively "infinite"; however, since the crystal presumably is in a non-catastrophic state, the integral $\int_{\tau_e} \cdots d\tau_e$ must converge.

Accordingly,

$$\frac{1}{|\mathbf{r}_n - \mathbf{r}_e|} = 4\pi \sum_{\ell=0}^{\infty} \sum_{m=-\ell}^{\ell} \frac{1}{2\ell+1} \frac{r_n^{\ell}}{r_e^{\ell+1}} Y_\ell^m(\theta_n, \phi_n) Y_\ell^{-m}(\theta_e, \phi_e)$$

where $Y_\ell^m(\theta_i, \phi_i)$ are the normalized spherical harmonics in the i=n,e spaces. We can then write Eq.(2) as

$$W = \sum_{\ell} \sum_{m=-\ell}^{\ell} N_\ell^m E_\ell^{-m} \tag{3}$$

where

$$N_\ell^m = \sqrt{\frac{4\pi}{2\ell+1}} \int \rho_n(\mathbf{r}_n) r_n^{\ell} Y_\ell^m(\theta_n, \phi_n) d\tau_n$$

$$E_\ell^{-m} = \sqrt{\frac{4\pi}{2\ell+1}} \int \rho(\mathbf{r}_e) r_e^{-(\ell+1)} Y_\ell^{-m}(\theta_e, \phi_e) d\tau_e .$$

The element $\rho_n d\tau_n$ is equivalent to $\Psi^* \Psi d\tau_1 \cdots d\tau_A$ where $\Psi = \Psi(R_1, R_2 \cdots R_A)$ describes the state of the nucleus. With respect to this wavefunction Ψ, N_ℓ^m then corresponds to the expectation value of the operator N_ℓ^m:

$$N_\ell^m = \sqrt{\frac{4\pi}{2\ell+1}} \sum_{j=1}^{A} e_j R_j^{\ell} Y_\ell^m(\Theta_j, \Phi_j), \tag{4}$$

$e_j = +e, 0$ for a proton, neutron respectively. For the case in which the "external" charges are external to the nucleus and its associated electrons, we similarly can set E_ℓ^{-m} equal to the expectation value of the operator E_ℓ^{-m}:

$$E_\ell^{-m} = \sqrt{\frac{4\pi}{2\ell+1}} \sum_{k} e_k r_k^{-(\ell+1)} Y_\ell^{-m}(\theta_k, \phi_k) . \tag{5}$$

In this case ρ_e corresponds to a wavefunction ψ_e specifying the charge density distribution. In the simplest version ψ_e is a "comb" function corresponding to an assembly of point ions specified by the index k. If, in fact, the "external" charges refer to the electronic charges associated with the

reference nucleus Eqs.(3) and (5) still remain valid; in
this case the ψ_e correspond to the usual electronic wave-
functions and the e_k are all equal to -e. In either case
H, the Hamiltonian operator corresponding to W, is:

$$H = \sum_m \sum_\ell N_\ell^m E_\ell^{-m} . \tag{6}$$

By definition N_ℓ^m, E_ℓ^{-m} are tensor operators of order ℓ. We
call N_ℓ^m the nuclear multipole moment of order ℓ. Relation
(6) can be simplified by invoking the symmetry of ψ which is
either even or odd with respect to inversion. If P denotes
the parity (inversion) operation, we have $\underline{P}N_\ell^m = N_\ell^m$,
$\underline{P}(\psi*\psi) = +\psi*\psi$ and $\underline{P}Y_\ell^m = (-1)^\ell Y_\ell^m$; hence only the even nuclear
multipole moments are observable. The first, $\ell=0$, is the
monopole term; in its simplest form this portion of W corresponds to the Madelung energy. The next, $\ell=2$, is the quadrupole term W_Q and is of immediate interest to us;

$$W_Q = \sum_m N_2^m E_2^{-m} \tag{7}$$

or in operator form,

$$H_Q = \sum_m N_2^m E_2^{-m}. \tag{8}$$

From the previous definitions (Eq.(3)) we find for the E_2
components (or equivalently the E_2 operators):

$$E_2^0 = \frac{1}{2} \sum_k e_k \frac{(3z_k^2 - r_k^2)}{r_k^5} = -\frac{e}{2} \left(\frac{\partial^2 V}{\partial z^2}\right)_{r=0} \equiv -\frac{e}{2} V_{zz}$$

$$E_2^{\pm 1} = \frac{\sqrt{6}}{2} \sum_k e_k z_k \frac{(x_k \pm iy_k)}{r_k^5} = -\frac{e}{\sqrt{6}} (V_{xz} \pm iV_{yz}) \tag{9}$$

$$E_2^{\pm 2} = \frac{\sqrt{6}}{4} \sum_k e_k \frac{(x_k^2 - y_k^2 \pm 2ix_k y_k)}{r_k^5} = -\frac{e}{2\sqrt{6}} (V_{xx} - V_{yy} \pm i2V_{xy}).$$

In the above, V is the potential of the external charge distribution seen at the nucleus; the V_{ij} are the appropriate
second-order derivatives; the convention is such that the

electronic charge -e is factored out of the V expressions.

The N_2 operator components are developed in terms of relations analogous to (9). With the aid of the Wigner-Eckart theorem the operators can be written in terms of the nuclear spin operator $\underset{\sim}{I}$ as follows:

$$N_2^0 \equiv Q^0 = \frac{eQ}{I(2I-1)} \cdot \frac{1}{2}[3I_{\sim z}^2 - I^2]$$

$$N_2^{\pm 1} \equiv Q^{\pm 1} = \frac{eQ}{I(2I-1)} \cdot \frac{\sqrt{6}}{4}[I_{\sim z}I_\pm + I_\pm I_z] \qquad (10)$$

$$N_2^{\pm 2} \equiv Q^{\pm 2} = \frac{eQ}{I(2I-1)} \frac{\sqrt{6}}{4} I_{\sim \pm}^2$$

where

$$eQ \equiv \int \rho_n r_n^2 (3\cos^2\theta_{n,I} - 1) d\tau_n.$$

($\theta_{n,I}$ is measured with respect to the nuclear spin direction.) Also $I \geq 1$ for these results to be meaningful.

Equations (7) and (8) can now be rewritten as

$$W_Q = \frac{-e^2}{6} \sum_{i,j} Q_{ij} V_{ij}; \qquad H_Q = \frac{-e^2}{6} \sum_{i,j} Q_{ij} \frac{\partial^2 V}{\partial x_i \partial x_j}$$

in energy and operator form respectively. Both Q_{ij} and $\underset{\sim}{V}_{ij}$ are symmetric second rank tensors. Since the off-diagonal elements of $\underset{\sim}{V}_{ij}$ are symmetric, the 9 components reduce to 6 independent quantities. If necessary, by transforming the coordinate system (x y z) appropriately into a new one (x' y' z') we can set any non-zero off-diagonal components to zero, leaving 3 diagonal components. (For convenience we will drop the primes on the coordinate designations.) Finally, from Laplace's Equation $\sum_i V_{ii}=0$ hence just 2 independent quantities remain. It has become conventional to label the axes of our coordinate system such that $|V_{xx}|<|V_{yy}|<|V_{zz}|$.
The EFG then is expressed in terms of two parameters $q \equiv V_{zz}$ and $\eta \equiv (V_{xx}-V_{yy})/V_{zz}$. An η value of zero corresponds to axial symmetry about the z-direction. The details of the determination of η and e^2qQ from nuclear quadrupole resonance (or more particularly Mössbauer Effect) spectroscopy do not interest us here.[2] We are concerned with the calculation of

q and η.

II. CALCULATIONS OF THE ELECTRIC FIELD GRADIENT (EFG)

The conceptual problems that arise in the calculation of the EFG components comprise a small portion of what is called "Ligand-Field Theory" or "Crystal-Field Theory." The various approaches used in these EFG calculations thus reflect correspondingly different procedures in Ligand-Field Theory. The Ligand-Field theorists usually are concerned with electronic rather than nuclear energy level differences; however, it should be noted for example that the parameter which the ligand-field theorists call B_2^0 (or B_0^2, A_2^0) corresponds to "q"; the ligand-field shielding coefficient σ_2 is closely related to the shielding coefficients arising in nuclear quadrupole resonance.

A. Shielding Effects

The wavefunctions corresponding to the electronic energy states of an ionic system set in a crystal lattice are specified by the stereochemistry - the scientific discipline pertinent to these considerations is called group theory. (We will discuss these matters still within the framework of our "ionic" model as defined previously.) The wavefunctions are formed of appropriate combinations of the free-ion wavefunctions; presumably these are of the Hartree-Fock type. However, these resultant wavefunction combinations are inadequate when considering the effect of the "external" environment on the nucleus - we must allow for the distension of the electronic shells by the ambient; hence "shielding."

1. Closed-Shell Ions. The simplest situation ensues in ions having a closed-shell electronic structure (an atomic S state, an A or A_1 state in the jargon of group theory). A classic example is Fe^{3+} [Ar$3d^5$].   We would not expect an EFG contribution from the closed shell; however, the field associated with the "lattice" (the external environment) polarizes this shell leading to an additional contribution to the EFG; the fractional correction is expressed as the Sternheimer shielding factor, γ_∞. If we now denote the original unshielded EFG value by $q'_{lattice}$, we have

$$q_{lattice} = (1 - \gamma_\infty)q'_{lattice} .\qquad (11)$$

From Eq.(9), $q'_{lattice} = -\frac{1}{e} \sum_k e_k (\frac{3z_k^2 - r_k^2}{r_k^5})$. The relation
is presumed to hold for the xx and yy EFG components as well,
i.e. η is unchanged.

Such shielding effects have been calculated by Sternheimer,[3] Das and Bersohn,[4] Watson and Freeman[5] and others.
The shielding factors for atoms with $A \geq 16$ generally are
negative, leading to "antishielding" or enhancement. The
γ_∞ value for Fe^{3+} is -9.18, for example.

2. Incompletely Closed-Shell Ions. The "valence" electrons in such structures do contribute to the EFG; we denote
this contribution by $q'_{valence}$. Moreover, the fields due to
these valence electrons polarize the inner spherical electron core (the outer as well, in the case of f-electrons)
leading to additional EFG contributions. Calculations have
been made by the methods of Sternheimer and Watson and Freeman. The result is expressed in the form

$$q_{valence} = q'_{valence} (1-R) \qquad (12)$$

where, usually, $0<R<1$. From Eq.(9) $q'_{valence} =$

$\sum_i \int \psi_i^* \frac{3\cos^2\theta_i - 1}{r_i^3} \psi_i d\tau_i$ where $\sum_i$ runs over the valence elec-

trons. Finally,

$$q = q'_{valence} (1-R) + q'_{lattice} (1-\gamma_\infty). \qquad (13)$$

It should be remembered that the derivations leading up to
Eq.(13) presume that we can define an "outside" (lattice
charges and their associated electrons) and an "inside"
(valence and core electrons) as far as charge distributions
are concerned. This is consistent with the ideal "ionic"
model. It may be an oversimplification when other approaches
are used.

B. EFG Calculations in Ionic Crystals

EFG calculations appropriate to sites in ionic crystals
(lattice-sum calculations) have been made with increasing
facility in recent years. The modern high speed computer
has relieved much of the tedium. The ingenious use of appropriate groupings or "clusters" of ions has speeded up

convergence enormously. Most of the computational uncertainties formerly persistent in many such calculations have
been removed. (Of course non-geometrical parameters remain
which are imperfectly known both theoretically and experimentally.) However, it must be remembered that the mathematical validity of these results has nothing to do with
their physical validity. Indeed the removal of mathematical
ambiguity has made scientific criticism of the lattice-sum
method increasingly meaningful. Nevertheless, the lattice-
sum approach is the only method presently available for computing the electric-field components of a multi-body system
(crystal) in any convenient form. It provides a convenient
reference with which to compare the calculations made using
more sophisticated models. And even when a molecular-orbital
method is used instead of the ionic model to treat the central ion and its immediate environment, the effect of the
remainder of the crystal is usually expressed via the lattice-
sum approach; nothing else is currently available for this
purpose. We now proceed with an exposé of the lattice-sum
method.

    In the simplest approximation we represent the crystal
as an array of point charges (monopoles). Summation of the
desired quantities (just V_{xx}, V_{yy}, V_{zz} in the simplest case)
proceeds in a straightforward manner. In some cases it may
be possible to handle the problem algebraically; however,
more usually, positions of the ions are specified by "special
position parameters" and a detailed arithmetical computation
is required. Symmetry relations, which are frequently obvious, reduce the number of quantities whose actual calculation
is required.

    From exposure to texts on electrostatics, one might
assume that convergence is assured by employing spherical
domains of increasing radii. This is not the case; the net
included charge within a sphere is not zero in a practical
case, and the calculators (human and machine) are plagued
with deBoer (surface) effects. However a few years ago,
Wood[6] pointed out the utility of a simple highly convergent
method of calculating Madelung constants in crystals of low
symmetry. Although this in some ways is a simple modification of the Evjen method[7] the procedure proposed by Wood
seemed to have been overlooked previously (and since[8]). In
his method ions are grouped into cells or "clusters" of zero
electrical charge. The clusters were stacked about the

central cluster containing the reference ion. (The cells used by Wood in his example were prismatic in form.) This method eliminates surface effects and permits rapid convergence. Wood found the rapidity of convergence to be dependent upon the method used to stack the cells. Before discussing the application of the chargeless cluster method to some cases of topical interest we will consider some elaborations of this simple lattice-sum method.

The next stage of sophistication in the lattice-sum approach involves consideration of dipoles. Ions which are not located at centers of inversion will be subject to non-zero monopolar electric fields and polarize. These ionic dipoles in turn will set up polarizing fields (the "reaction" fields). If the ionic polarizibilities are specified, the dipole values can be determined in a self-consistent manner:

$$\vec{P} = (\underset{\sim}{1} - \underset{\sim}{\alpha} \underset{\sim}{K})^{-1} \underset{\sim}{\alpha} \vec{E} \quad . \tag{14}$$

Here $\vec{P}$ is the dipole hypervector, $\underset{\sim}{\alpha}$ the polarizibility hypertensor matrix, $\underset{\sim}{K}$ the reaction hyperfield matrix, and $\vec{E}$ the monopolar electric field hypervector. We calculate $\vec{E}$ from the monopolar charge distribution, compute $\underset{\sim}{K}$ which is essentially a geometrical parameter, and supply values of $\underset{\sim}{\alpha}$.

The selection of the α-components is not non-trivial. It has been customary to associate $\underset{\sim}{\alpha}$ with "electronic polarizibility." The correlation between microscopic and macroscopic polarization effects in crystals has been discussed by Tessman et al.[9] and by Bolton et al.[10] If we use this model in its simplest form we find a molecular polarizibility α_m by application of the Clausius-Mossotti relation:

$$\frac{4\pi}{3} \frac{\alpha_m}{V} = \frac{\mu^2 - 1}{\mu^2 + 2} \quad . \tag{15}$$

Here V is the molecular "volume" and μ the optical index of refraction. From a systematic examination of various compounds Tessman et al. were able to deduce "individual" electronic polarizibilities for a number of common ions. However there are many difficulties of a systemic nature in such an analysis. From inspection of their data it is clear that one can not generally deduce an α_m from a simple addition of "known" atomic α's. The α_m trends found in particular crystal systems reflect, presumably, degrees of departure from an "ionic" model. Theoretical calculations of α are not of

much use here since they apply to only relatively simple
physical systems. From a pragmatic point of view probably
the best that can be done is to establish a range of α values
from available refractive data and to hope that the dipole
contribution to the lattice sum q' is relatively small. (An
interesting analysis in the BeO system which distinguishes
between several different dipolar polarizibilities has been
made by Taylor and Das.[11])

The relations appearing in Eq.(14) in practice may be
much less formidable than would first appear. For example,
cation polarizibility is usually neglected. Application of
hyperspace geometry frequently is relatively simple. We will
examine the α-Fe_2O_3 structure in this regard. As indicated
in Fig. 1 we can consider the α-Fe_2O_3 crystal as being composed of a distribution of molecular Fe_2O_3 bipyramids. The
lines running through the two Fe^{3+} ions of each bipyramid are
directed parallel to the hexagonal crystal axis ($\underline{C_o}$). The O^{2-}
ions are disposed on the vertices of the equilateral triangle
forming the medial plane of the bipyramid. The Fe dipoles

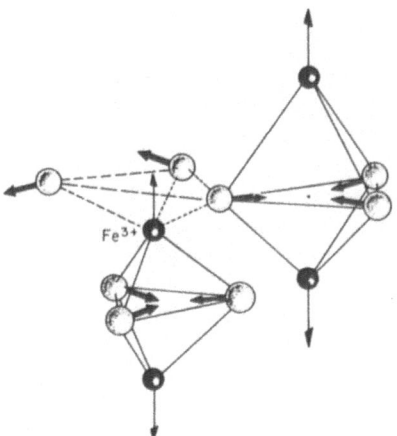

Fig. 1. Two of the bipyramids of the α-Fe_2O_3 lattice. The
black spheres represent Fe^{3+} ions; the white spheres represent O^{2-} ions. The six nearest neighbor O^{2-} ions about one
of the Fe^{3+} ions are depicted. Three of these oxygens are
associated with the bipyramid of the reference Fe^{3+} ion; the
remainder are associated one each with the three bipyramids
immediately above the reference Fe^{3+} ion. Molecular orbital
calculations are made for a $(FeO_6)^{9-}$ complex. Lattice sum
calculations are made most convergently by considering contributions from Fe_2O_3 bipyramids.

must all have the same value and are alternately parallel
and anti-parallel to the $\underline{C}_o$ axis. The O dipoles are all
equivalent in magnitude also; in each bipyramid the oxygen
P-vectors are directed inwardly toward the center of the
triangle. In this case, Eq.(14) can be written using just
2-component vectors (P_O, P_{Fe}; E_O, E_{Fe}) a simple α tensor
($^\alpha\alpha_{Fe}$) and a 2x2 $\underline{K}$ matrix. The $\underline{K}$ matrix components are
not all independent,[12] simplifying the calculations still
further. (Unfortunately, in other cases the coordinate system for the hypervectors and tensors may involve different
directions as well as different constituents.) From a computational point of view, the biggest problem in the dipole
contributions had been the evaluation of the slowly convergent monopolar electric fields. The cluster method makes
this trivial. After superimposing the dipole lattice mathematically upon the monopolar lattice, the $\underline{K}$ matrix and unit
dipolar tensor q'_{dip} are evaluated. The dipolar EFG contribution is $q'_{dip}\vec{P}$. In final form the (unshielded) lattice sum
EFG is

$$\vec{q}'_{lattice} = \vec{q}'_{monopole\ lattice} + q'_{dipole\ lattice}\vec{P}. \quad (16)$$

where, as explained before, the hypervectors and tensors are
defined with respect to unique sites as well as to coordi-
nates in conventional geometry.

The next escalation in the lattice-sum calculation involves ionic quadrupolar polarizibility effects. Unfortunately quadrupolar polarizibility tensors are not very well
defined either experimentally or theoretically. Unless sim-
plifications are made in the mathematics, complicated reaction effects and matrices occur. Quadrupolar polarizibility[12] has been used to eliminate some of the anomalies arising in lattice-sum EFG calculations; whether this procedure
is justified or not has been a matter of dispute.

Before proceeding further it is well to re-emphasize
the distinction between the mathematical and physical validity of these lattice sum calculations. Current calculation
methods guarantee mathematical validity. The question of
physical validity is more obscure; for example we might expect the successive multipolar contributions to the EFG to
contribute decreasingly but the trend in the few orders calculated in any particular example may not be meaningful. A
question of philosophy also arises here which transcends the

physical model to be parametricized. The lattice-sum method
emphasizes the space group characteristics of the crystal.
A simple molecular orbital calculation ignores the space
group (except possibly insofar as the contributions of the
more distant elements of the crystal are concerned) but emphasized the symmetries of the point group characterizing the
molecular complex under analysis.

In this connection calculations have been made using an
equivalent of a "poor man's" parametricized molecular-orbital
method. Dipoles were assigned to the nearest-neighbors of
the reference ion and directed along the bond axes. A more
up-to-date version of the lattice sum extension of such a
calculation might be the analysis of the complex of interest
by molecular-orbital methods, the assignment and addition of
equivalent dipoles to the nearest neighbor ions, the genera-
tion of the entire lattice from these constituents, and the
evaluation of the distant contributions to the reference ion
by lattice-sum methods. Calculations involving shifting of
charge from one ionic constituent to another (the amount of
shift possibly being determined from a molecular orbital calculation) have also been made. Fractional charges have also
been placed at strategic positions on bonds in ad hoc attempts
at generating desired lattice sum values.

We have not discussed (for good reason!) a conceivable
simple brute force alternative to our considerations. Dis-
regarding the computational labor involved, the EFG could be
obtained from the electron density distribution in the lattice
by numerical integration. Unfortunately, crystallographers for
the last decade or so have not been interested in producing
these electron density plots. The considerable effort required (both human and computational) does not appear justified since the obtainable accuracy is limited by experimental
difficulties. The electrons of primary concern to us are the
outer "valence" electrons. The relevant information is contained mainly in the low order reflections of the diffraction
pattern. As a result of difficulties with extinction, the
intensities of these reflections cannot be determined very
well; even under favorable circumstances uncertainties in the
electron distribution occur which correspond to 0.5-1.0
electron.

Before closing this section we mention also the more
rapidly convergent lattice-sum procedure advocated by Nijboer

and deWette;[13] this involves in part summation over the in-
verse lattice. Regardless of the merits of this procedure,
we believe that the efficacy of the various cluster methods
has made resort to such mathematically elegant analyses un-
necessary from a practical point of view.

C.   Speculations on Substituents

The problem of substituents is a subtle one; the unini-
tiated can be misled easily by generalization from a particular example. There is no reason for the immediate neighbors
of an ion to retain their positions upon its substitution by
an alien species. These "relaxation" effects will affect the
EFG components of course. Within the framework of the lattice-
sum method, for example, the modification in the calculations
could be made fairly simply if the local distortions were
known and the dipole contributions weren't too large. (We
assume that the number of substituents is sufficiently small
to permit us to ignore the changed contributions of all the
other relaxed environments in the crystal to our sum.) However, determination of those local distortions by x-radiation
techniques is a very difficult task.[14] Also the relation
between the EFG's observed in "100%" crystals and substituted
hosts is not at all obvious.

We illustrate some of these problems by considering substitution of Al^{3+} by Fe^{3+} in two different hosts; we recol-
lect that Fe^{3+} is larger than Al^{3+}, the ionic radii being
0.64 and 0.45 Å respectively. We consider the following two
examples: the octahedrally-coordinated Al site in α-Al_2O_3
and the tetrahedrally-coordinated Al site in $LiAl_5O_8$. The
ΔE_Q reported for Fe^{57m} in α-Fe_2O_3 was[15] 0.440 mm/sec. The ΔE_Q
reported for Fe^{57m} in α-Al_2O_3 was ten percent larger.[16] On
the other hand the ΔE_Q reported for Fe^{57m} on the tetrahedral
A sites in $LiAl_{5-0.25}Fe_{0.25}O_8$ was[17] 0.675 mm/sec. But the
corresponding splitting in $LiFe_5O_8$ was estimated not to ex-
ceed[17] 0.02 ± 0.06 mm/sec. In each crystal type the EFG computed by the lattice-sum method was very similar for both the
dilute and concentrated material. It is very hard to deduce
a systematic trend from data such as these. In Fig. 2 we depict
the variation of the monopolar contribution to q' in α-Al_2O_3
when the reference ion is shifted along the C_0 axis. The
positions of the remaining ions are assumed to be fixed. The
small oxygen triangle is the one in the medial plane.

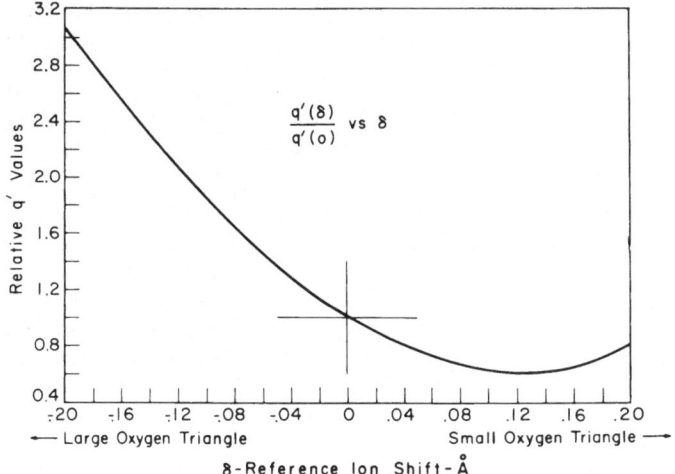

Fig. 2. Variation of the monopolar contribution to q' in α-Al$_2$O$_3$ with respect to shifts in reference ion position along the C$_o$ axis. Unpublished data from J. O. Artman and J. C. Murphy and from N. Laurance.

D. Three Selected Crystal Systems

1. α-Fe$_2$O$_3$ Space Group R$\bar{3}$c (D$_{3d}^6$). The most convenient chargeless cluster in α-Fe$_2$O$_3$ is the bipyramidal molecular unit which we had discussed previously. See Fig. 1. The Fe-O distances in these bipyramids are 2.116 Å. The distances between an Fe^{3+} ion and its other three nearest O^{2-} neighbors are 1.945 Å. Although some attempt was made to determine optimum stacking procedures, the calculations in practice were performed over studded spheres centered about the centroid of of the cluster containing the reference Fe^{3+} ion. Only clusters whose centroids were within any given sphere radius were admitted to the calculation. The calculations proceeded rapidly for two reasons: 1) the machine logic required to generate acceptable points was much simpler and quicker than that of the conventional spherical method; 2) the computations of course converged much more rapidly. When necessary, off-center effects were averaged out by considering, for example, the Fe ion to be alternately in the "up" and "down" position. Since the α-Fe$_2$O$_3$ lattice is hexagonal, possibly still better convergence could have been obtained by using ellipsoids of revolution with axis ratios proportional to the crystal C$_o$/A$_o$ ratio; since, however, machine times and costs were relatively insignificant this matter was not pursued. Spheres of 4A$_o$ in

radius ($\underline{A}_0$ = 5.035 Å) were adequate for all purposes. The monopolar and dipolar contributions to q' were[15] +.08811 and +.01413 respectively in units of (Å)$^{-3}$. The Q(Fe^{57m}) value that followed from these data was +0.283b. The sensitivity of this calculation to α-Fe$_2$O$_3$ structural parameters was noted.[15] Details of the cluster calculation (including convergence comparisons with more conventional lattice sum procedures) were given in earlier papers[18,19] and will not be repeated here. The modification of a calculation of this type by the inclusion of quadrupolar electrical polarizibility is discussed in Ref. 12. We note also earlier calculations of the EFG in α-Al$_2$O$_3$ and spinel (MgAl$_2$O$_4$) by Brun and Hafner.[20] These authors considered also the electrical dipolar and quadrupolar polarizibility contributions to the EFG. However the lattice-sum calculations were made by slowly convergent techniques.

2. MgAl$_2$O$_4$ and its Isomorphs (Spinel)

Space Group F4$_1$/d$\bar{3}$2m(O$_h^7$). In this case the structure is generated by translation of a cubic primitive cell containing 8 formula units. Unlike α-Fe$_2$O$_3$ where all the Fe sites are equivalent as far as quadrupole effects are concerned, the spinel structure has two types of sites (A-tetrahedral, and

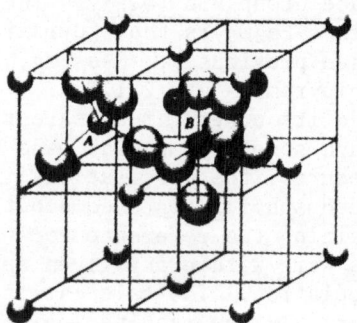

Fig. 3. The unit cell of the spinel lattice. Not all of the ions are shown. The length of a cube edge is about 8 Å. The large spheres represent O^{2-} ions, the small light spheres represent ions in tetrahedral (A-type) sites, the small dark spheres represent ions in octahedral (B-type) sites. In "normal" spinel these sites are occupied by divalent and trivalent cations respectively. (From A. H. Morrish, The Physical Principles of Magnetism, Wiley, New York, 1965.)

B-octahedral) potentially available for occupancy by Fe ions.
The structure is depicted in Fig. 2. A fairly general computer program appropriate to orthorhombic coordinate systems
was set up. (It was applied later to the FeOCl lattice.) A
3-dimensional gridwork was constructed by dividing each edge
of the primitive cell into eighths. The ions were located;
using special position parameters as necessary, with respect
to the nodes of this grid. The node corresponding to the desired reference ion was set at a corner of the array by trans-
lating layers appropriately. The re-arranged cube was then
symmetrized by halving ions located at cube interfaces, quart-
ering ions located on cube edges, "eighthing" ions located at
cube corners and re-distributing the sliced-off pieces appropriately. This symmetrization was found to be essential for
rapid convergence. Summations were performed over arrays of
these symmetrized cubes whose centers were contained within
spheres of increasing radius. (In the more general orthorhombic case ellipsoids with principal axes proportional to
the $\underline{a}$, $\underline{b}$, $\underline{c}$ lattice constants were used.) No attempt was made
to determine optimum stacking. As before convergence was rapid;
computation time was minimal. Radii of 4 or 5 units were sufficient for convergence.

    Comparatively little quadrupole-splitting data are avail-
able from Mössbauer Effect experiments in the fully substituted
spinels ($Al^{3+} \rightarrow Fe^{3+}$, $Mg^{2+} \rightarrow$ divalent ion) or "ferrites."[21] A
A paper by Hudson and Whitfield (designated hereafter as
H&W),[22] compares the splittings found in $ZnFe_2O_4$ and $CdFe_2O_4$
with deductions from lattice-sum EFG calculations. In standard spinels there is no EFG at the A-sites. The EFG tensor
at the B-sites is axially symmetric with the various principal axes along <111>. The O^{2-} ions experience electric fields,
all of the same magnitude, along <111>. H&W found the calculated EFG to be very sensitive to the value of $\underline{u}$, the special
position parameter for the oxygen array, which ideally is 3/8.
In many instances the value of $\underline{u}$ may not be known too well.
They found the dipolar contribution to the EFG to be substantial in a formal computational sense. However, in order to
make the $\underline{q}'$ calculations fit the experimental data, they suggest that the "effect of the dipolar terms is to a large extent cancelled by the effects of covalency and induced electric quadrupole moments." The covalency effect cited by
these authors is associated by them primarily with the tightly
bound ions at the tetrahedral A-sites. The implication is
that the charge distributions associated with these "covalent"
sites can not be properly (or easily) represented in terms of

the monopole-dipole model. Thus the EFG calculations for the relatively "ionic", octahedrally coordinated B-sites made using this model are not too pertinent. Unfortunately there is a nagging suspicion that these indirect "covalency" effects occur also in other systems. (For example, it was found when making EFG calculations pertinent to the Fe^{3+} sites in $CuFeO_2$ and $AgFeO_2$ that the dipole contributions to $\underline{q}'$ were very large in a formal sense; the resultant $\underline{q}'$ values were much too large.[23])

   H&W unfortunately made their lattice sum calculations by conventional techniques. Since the convergence was poor it was felt advisable to redo these calculations by the symmetrized chargeless cluster method. These results were reported at an American Physical Society meeting;[24] we depict some figures from this talk below. In Fig. 4 we plot the monopolar electric field at an oxygen site in spinel for a $\Delta(\Delta\equiv u-3/8)$ of 0.012. For comparison with H&W, computations were made within spheres up to radii of 10 lattice constants. The

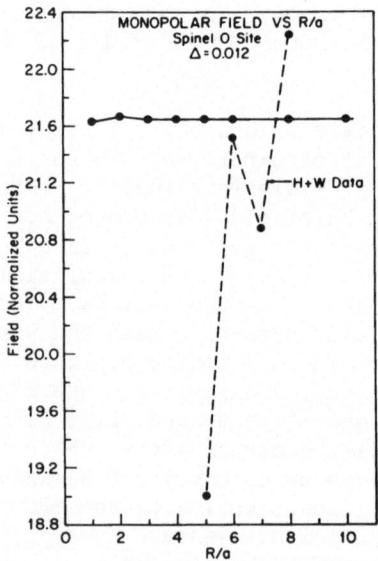

Fig. 4. The monopolar electric field at an oxygen site in spinel versus computation sphere radius. (The normalized units correspond to the charge $\underline{e}$ and the cube edge $\underline{a}$ being set equal to 1.) The value of Δ is 0.012 ($\Delta\equiv u-3/8$ where 3/8 is the "ideal" oxygen lattice special position parameter.) These data were obtained using the symmetrized chargeless cluster method. Some data obtained by conventional means by H&W[22] are shown for comparison.

superiority of the cluster method is clear. In Fig. 5 we plot
the monopolar electric field at a spinel oxygen site as a function of Δ. Fortunately (fortuitously?) the "average" electric
field values selected by H&W corresponded closely to the cluster method results shown in these latter two figures. (See
Fig. 2 of Ref. 22 for comparison.) In Fig. 6, taking z in the
[111] direction, we plot the monopolar V_{zz} value at a spinel
B-site for a wide range of Δ excursions. (An equivalent plot
over a more restricted Δ range can be found in Fig. 4 of Ref.
22.) The final results of this effort for spinel geometry
reproduced those of H&W over that part of the Δ range common
to both calculations.

In view of the arguments above relating to the inadequacy
of the lattice-sum approach, the reasons for some of the
caveats on inferring conclusions from substituents become
clearer. In this connection one may cite the recent paper
by Rosenberg et al.[25] on the NMR of Al^{27} in a number of spinels and on the Mössbauer spectra of Fe^{57m} as a substituent
in two of these. Using the Al^{27} data in $CoAl_2O_4$ (a normal
spinel) as a reference, these authors obtain a $Q(Fe^{57m})$ value
of $0.20 \pm 0.01b$ from data on $CoAl_2O_4$:Fe! To argue from this[25]
that the door has been opened, therefore, for a meaningful
comparison of NMR and Mössbauer data in various lattices is a
bit simplistic.

The $LiFe_5O_8$ structure[26] is an interesting variation of
the spinel lattice. It exists potentially either in the
ordered or disordered form. In the ordered phase the space
group is $P4_332(O^7)$ or the enantiomorph $P4_132(O^6)$. There are
four formula units per cubic primitive cell. In this structure the Fe^{3+} ions occupy both A- and B-sites. The EFG is
non-zero at both types of sites; the A-site EFG is axially
symmetric along <111> directions; the B-site has just a 2-
fold symmetry axis, a non-zero η is required here. The 32
oxygen ions are divided into two groups: 8 experience electric fields along <111> directions; the remaining 24 experience skewed electric fields. EFG calculations had been computed for the Al isomorph.[27] Since these obviously were
crude in certain respects, a chargeless cluster calculation[24]
was performed. This system demanded an interesting journey
into hyperspace! The Mössbauer results[17] in $LiFe_5O_8$ and
$LiAl_5O_8$:Fe had been mentioned previously in this paper. The
site of primary interest here was the tetrahedral A-site.
Unfortunately the dipolar contribution to the EFG was much

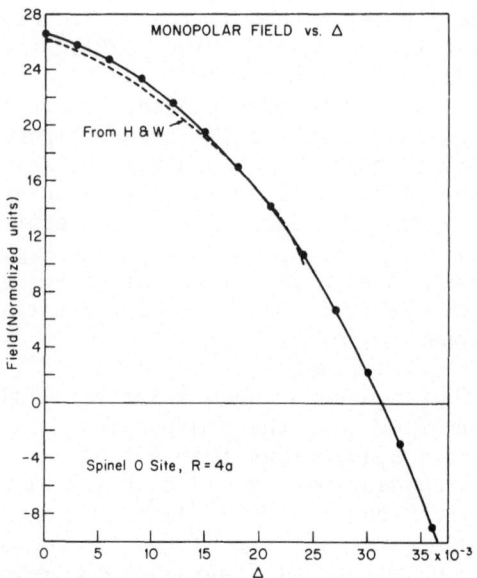

<u>Fig. 5.</u>   Monopolar electric field at an oxygen site in
spinel versus Δ. Normalized units are used. Data from H&W[22]
are shown for comparison.

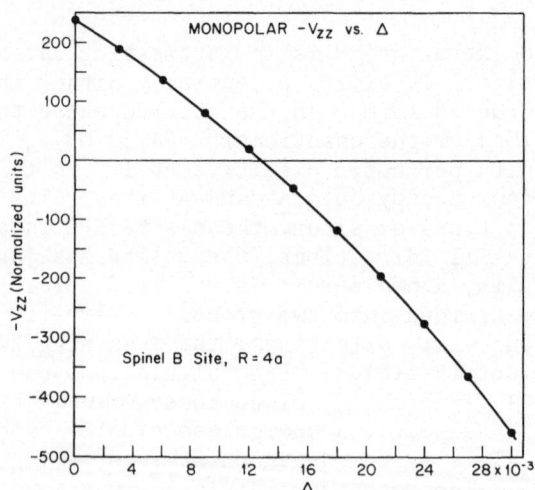

<u>Fig. 6.</u> The monopolar electric field gradient at the octahe-
dral B site in spinel versus Δ.

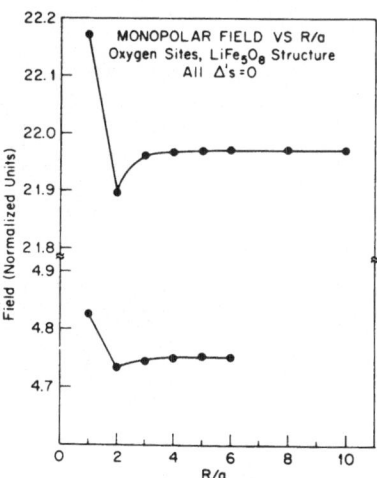

<u>Fig. 7</u>. Monopolar electric field versus computation sphere radius at the two types of oxygen sites in LiFe$_5$O$_8$. The data for the "skewed" oxygen sites lie above. The rapid convergence of the symmetrized chargeless cluster method is clear. The results are expressed in normalized units.

too large to be meaningful. As an example of these computations we indicate in Fig. 7 the nature of the convergence of the monopolar electric field at the two types of oxygen sites.

Those interested in a heroic computation (by conventional methods) of the EFG in a still more complicated system would do well to examine the paper on chrysoberyl, (BeAl)$_2$SiO$_4$, by Rao and Rao.[28]

3. <u>FeOCl Space Group: Pmnm(D$_{2h}^{13}$)</u>. This orthorhombic crystal has been the subject of recent crystallographic and ME investigations and of EFG calculations.[29] The structure is shown in Fig. 8. There are two FeOCl formula units per unit cell. All ions are on two-fold special positions with point symmetry mm(C$_{2v}$). The principal axes of the EFG tensor, (which is non-axially symmetric), are required to be parallel to the <u>a</u>, <u>b</u>, and <u>c</u> crystallographic axes. The lattice constants are respectively 3.780, 7.917 and 3.302 Å. Symmetry considerations require the EFG tensors for both Fe sites to be the same. Each Fe^{3+} ion is surrounded by a highly distorted octahedron. The edge lengths vary from 2.624 to 3.357 Å. With respect to the central Fe^{3+}, there are two O^{2-} ions at 1.964 Å, another two O^{2-} at 2.100 Å and two Cl$^-$ ions at 2.368 Å.

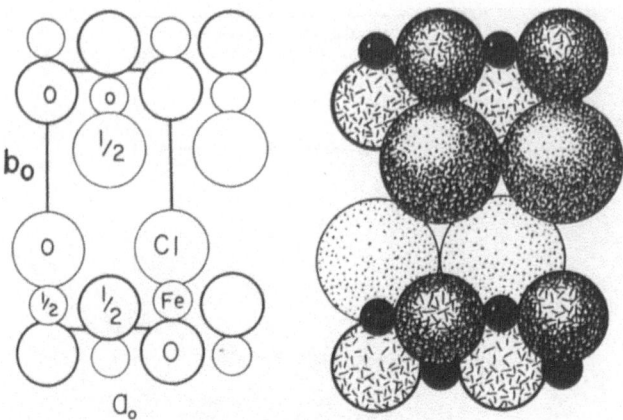

Fig. 8. The orthorhombic structure of FeOCl. At the left a projection along c with the origin in lower left. At the right, a packing drawing viewed along the c axis. The oxygen atoms are line-shaded, the chlorine atoms are dotted. Adapted from R. W. G. Wyckoff, Crystal Structures 2nd ed. Vol. 1 (Wiley, New York, 1963) p. 297.

FeOCl was selected for ME investigation as a crystal other than α-Fe_2O_3 suitable for checking model calculations. The observed ME splittings were large and the EFG calculations were not expected to be as sensitive to structural variations as in α-Fe_2O_3. In addition it was hoped that the nuclear quadrupole resonance of Cl^{35} and Cl^{37} could be observed. (These Cl resonances unfortunately still have not been detected in FeOCl.)

The EFG calculations were made in a routine fashion by the method described in the spinel case. As expected, the electric fields at all ion sites are directed either parallel or anti-parallel to the long (b) axis of the crystal. The two Fe^{3+} ions per unit cell in this sense are equivalent upon inversion (the O^{2-} and Cl^- ions are similarly pair-wise related). The dipole hypervector thus has only three components. The final calculations were simplified by setting α_{Fe} equal to zero. The experimental data to be fit were ΔE_Q=0.916 mm/sec, η=0.32. A reasonable fit to η was obtained by setting $\alpha_O = \alpha_{Cl}$=1 $\overset{\circ}{A}{}^3$. The $Q(Fe^{57m})$ calculated on this basis was 0.33b, which is much too large. We will discuss below some modifications of these EFG calculations by "overlap" contributions.

III.   A QUESTION OF Q!   DOES THE CURRENT FERROUS CONSENSUS
            CONTRAST WITH FERRIC PREDICTIONS?

In spite of, or rather partly because of, the removal of
computational uncertainties from the lattice sums it has become
evident that the $Q(Fe^{57m})$ data from ferrous and ferric compounds were still incompatible. The more recent $Q(Fe^{57m})$ determinations from ferrous data cluster about[30] 0.18b; the most
reliable ferric value, from the "classic" crystal α-Fe_2O_3, is[15]
0.28b. We had mentioned that this latter value can be scaled
downward by introduction of quadrupolar polarizibility, but
this is accomplished by arbitrarily adjusting an additional
parameter which is poorly defined.

A significant development in EFG calculation has occurred
within the last half year. This work has been pursued by
Sawatzky[31] and his associates and independently by Sharma.[32]
Sawatzky observed that a significant EFG contribution could
develop from the overlap distortion of the <u>filled</u> metal orbitals by the surrounding ligand orbitals. These authors seem
to have removed the discrepancy between the α-Fe_2O_3 and the
ferrous Q values. (Using the same procedure they have cleared
up also an inconsistency between the $Q(Al^{27})$ value determined
in α-Al_2O_3 and that calculated from atomic beam experiments.)
Within this framework, the dominant EFG contribution is associated with the <u>anti-bonding</u> orbitals.

These orbitals are written in the form:

$$\psi_j^k = \phi_j^k - \sum_i S_{i,j}^k \chi_i \qquad (17)$$

where ϕ_j^k is a free-ion 2p wavefunction pertinent to the kth
O^{2-} ion and the χ_i are Fe^{3+} wavefunctions. The overlap parameters, $S_{i,j}^k$, are defined in the usual way:

$$S_{i,j}^k = \int \chi_i^* \phi_j^k \, d\tau \quad .$$

Sawatzky determined that the principal contribution to $q'_{valence}$
(see Eq.(13)) from this mechanism is

$$\sum_{j,k} \sum_{i,i'} S_{i,j}^k S_{i',j}^k \int \chi_i^* (3\cos^2\theta_k - 1) r^{-3} \chi_i' d\tau \qquad (18)$$

where the χ_i represent Fe^{3+} 2p and 3p wavefunctions. Using a
shielding parameter, R, value of 0.32 and the α-Fe_2O_3 $q'_{lattice}$

value previously calculated[15] he found $Q(Fe^{57m})$ to be 0.17b.
Philosophically, this overlap mechanism corresponds to a distortion of the inner Argon core of the Fe^{3+} ion by the surrounding oxygen neighbors which is expressed through the $\bar{S}$
factors and $\cos\theta$ values. It thus is sensitive to geometrical
uncertainties.

Sharma's analysis is more general. He evaluates explicitly contributions to q' from the antibonding and bonding
oxygen orbitals. Since these are "distant" effects, their
contribution is multiplied by $1-\gamma_\infty$ instead of $1-R$. (Sharma's
$\alpha-Fe_2O_3$ results are expected to appear soon.[32]) In this analysis the nearest-neighbor contributions to the dipole lattice
sum are subtracted from the previously determined $q'_{lattice}$
value; they are now more explicitly and presumably more properly treated in terms of O^{2-} 2p wavefunctions.

We have mentioned in this paper recent ME analyses and
lattice-sum EFG fits pertinent to the non-axially symmetric
Fe^{3+} site in FeOCl. From the EFG point of view, this crystal
is relatively asymmetric, thus unlike $\alpha-Fe_2O_3$ small uncertainties in crystallographic parameters do not affect the EFG calculations appreciably. Unfortunately the fit of the conventional lattice-sum computations to the data was not very satisfactory. It would seem that such an asymmetric (but geometri-
cally still relatively simple) site would prove a more searching test of the apparently successful overlap approach. Sengupta and Artman[33] have made such an FeOCl calculation employing the simpler Sawatzky approach. The V_{xx} and V_{yy} contributions follow from Eq.(18) merely by replacing the $3\cos^2\theta_k-1$
term with $3\sin^2\theta_k\cos^2\phi_k-1$ and $3\sin^2\theta_k\sin^2\phi_k-1$ respectively.
The Cl^- contribution to the overlap was attributed to the Cl^-
3p orbitals. The O^{2-} contributions were treated as before.
This simple overlap calculation modified the previous FeOCl
fit substantially; however, it did not lead to anything like
the agreement achieved in $\alpha-Fe_2O_3$. In particular, the "reasonable" FeOCl EFG lattice sum fit mentioned previously, which had
matched the experimental η value of 0.32 but had corresponded
to a Q of 0.33b, was converted by the overlap contribution to
a η of 0.47 and a Q of 0.11b. More sophisticated studies of
overlap in the FeOCl system would prove intriguing. When the
dipole contributions of the nearest neighbors are subtracted
from the lattice-sums (as occurs in Sharma's overlap treatment)
almost all of the oxygen dipole contribution would be removed
and the chlorine dipole contribution would reduce considerably

The dipole contribution to the EFG in FeOCl is much greater than in α-Fe_2O_3. Thus a procedure which reduces the mathematical significance of such dipole contributions removes some of the uncertainties in FeOCl system calculations.

We have not alluded to the computational complexities of these overlap calculations. Various sequences of two- and three-center integrals must be evaluated, which the inexperienced face with dismay. Additional human and machine computer programs must be provided. It may be a bit too soon for some to accept the general reliability of the overlap method. Once again a question of scientific philosophy also must be raised. Parameters such as α, 1-γ_∞ and 1-R may be cynically said to represent the physicists' technique of avoiding chemistry. Such parameterization makes it possible for a wide variety of EFG calculations to be done comparatively simply and generally. However, the overlap method blurs the "inside", $(1$-$R)$, and "outside", $(1$-$\gamma_\infty)$, EFG shielding effects; should we go "all the way", dispense with these shielding factors, and do "good chemistry"?

IV.   IN CONCLUSION

We have intended in this paper to provide an eclectic rather than an encyclopedic review of EFG calculations in ionic lattices. However, the references cited are sufficiently diverse to provide an inquisitive reader detailed information on various aspects of the problem. We look forward in the future to the appearance of EFG analyses developed on less of a Procrustean basis.

V.   ACKNOWLEDGMENTS

The author wishes to acknowledge discussions with his Mellon Institute of Science colleagues F. deS. Barros, P. A. Flinn, M. Kaplan and W. Oosterhuis. The point-finding program of the lattice-sum computer programs was taken from a routine appropriate to corundum geometry which was prepared for D. S. McClure. This routine was kindly furnished to the author some years ago by H. A. Weakliem of the RCA Research Laboratories. The modifications of this routine for orthorhombic geometry

(including charge symmetrization) were made by J. P. Stampfel
and R. A. Heinz while they were students at Carnegie-Mellon
University. Many of the calculations in spinel-type geometries
were made by J. P. Stampfel. The typing of the final draft
was done by Kate F. Ellis.

Res Ipsa Loquitor

VI.   REFERENCES

1. For general reference material see M. H. Cohen and F. Reif,
    Solid State Phys. 5, 321 (1957), T. P. Das and E. L. Hahn,
    Solid State Phys. Suppl. 1, (1958).
    The analysis presented in this section parallels in some
    respects material found in pertinent chapters of E. A. C.
    Lucken, Nuclear Quadrupole Coupling Constants (Academic
    Press, New York, 1969). This text contains numerous
    typographical errors unfortunately.
2. For a discussion of this matter see the paper by R. L.
    Collins and J. C. Travis in Mössbauer Effect Methodology,
    Vol. 3, I. J. Gruverman, Editor (Plenum Press, New York,
    1967) p. 123.
3. R. M. Sternheimer, Phys. Rev. 130, 1423 (1963).
4. T. P. Das and R. Bersohn, Phys. Rev. 102, 733 (1956).
5. R. E. Watson and A. J. Freeman, Phys. Rev. 131, 250 (1963)
6. R. H. Wood, J. Chem. Phys. 32, 1690 (1960).
7. H. M. Evjen, Phys. Rev. 39, 675 (1932).
8. W. B. Bridgman, J. Chem. Ed. 46, 592 (1970).
9. J. R. Tessman, A. H. Kahn and W. Shockley, Phys. Rev. 92,
    890 (1953).
10. H. C. Bolton, W. Fawcett and I. D. C. Gurney, Proc. Phys.
    Soc. (London) 80, 199 (1962).
11. T. T. Taylor and T. P. Das, Phys. Rev. 133, A1327 (1964).
12. M. Raymond and S. S. Hafner, Phys. Rev. 1B, 979 (1970).
13. B. R. A. Nijboer and F. W. deWette, Physica 23, 309 (1957).
14. S. C. Moss and R. E. Newnham, Z. Krist. 120, 359 (1964),
    claim to have determined the Cr^{3+}-position in highly doped
    α-Al_2O_3:Cr^{3+} by x-radiation techniques.
15. J. O. Artman, A. H. Muir, Jr., and H. Wiedersich, Phys.
    Rev. 173, 337 (1968).
16. G. K. Wertheim and D. N. E. Buchanan in Proceedings of
    the Second International Conference on the Mössbauer
    Effect, Saclay, France, 1962 (Wiley, New York, 1962)
    p. 130.
17. F. deS. Barros, P. J. Viccaro and J. O. Artman, Phys.
    Letters 27A, 374 (1968). A comprehensive paper by

P. J. Viccaro, F. deS. Barros and W. Oosterhuis is in
preparation.

18. J. O. Artman and J. C. Murphy, Phys. Rev. 135, A1622
    (1964).

19. J. O. Artman, Phys. Rev. 143, 541 (1966).

20. E. Brun and S. Hafner, Z. Krist. 117, 63 (1962).

21. A review of the saturation magnetization and crystal
    chemistry of ferrimagnetic oxides has been written by
    E. W. Gorter, Philips Res. Rep. 9, 295; 321; 403 (1954).

22. A. Hudson and H. J. Whitfield, Mol. Phys. 12, 165 (1967).

23. A resumé of the $CuFeO_2$ EFG calculations was given in the
    meeting paper corresponding to A. H. Muir, Jr., H. Wied-
    ersich, R. W. Grant, and J. O. Artman, Bull. Am. Phys.
    Soc. 14, 134 (1969).

24. J. O. Artman, F. deS. Barros, J. Stampfel, J. Viccaro,
    and R. A. Heinz, Bull. Am. Phys. Soc. 13, 691 (1968).

25. M. Rosenberg, S. Mandache, H. Niculescu-Majewska, G.
    Filotti, and V. Gomolka, J. Appl. Phys. 41, 1114 (1970).

26. P. B. Braun, Nature 170, 1123 (1952).

27. G. H. Stauss, J. Chem. Phys. 40, 1988 (1964).

28. L. D. V. Rao and D. V. G. L. Rao, Phys. Rev. 160, 274
    (1967).

29. R. W. Grant, H. Wiedersich, R. M. Housley, G. P.
    Espinosa and J. O. Artman, Phys. Rev. B3, 678 (1971).

30. R. Ingalls, Phys. Rev. 188, 1045 (1969); J. Chappert,
    R. B. Frankel, A. Misetich and N. A. Blum, Phys. Rev.
    179, 578 (1969); H. R. Leider and D. N. Pipkorn, Phys.
    Rev. 165, 494 (1968); A. J. Nozik and M. Kaplan, Phys.
    Rev. 159, 273 (1967); C. E. Johnson, Proc. Phys. Soc.
    (London) 92, 748 (1967); F. S. Ham, Phys. Rev. 160,
    328 (1967).

31. G. A. Sawatzky and J. Hupkes, Phys. Rev. Letters 25,
    100 (1970); G. A. Sawatzky, F. van der Woude and J.
    Hupkes, J. de Physique (1971) (to be published).

32. R. R. Sharma, Phys. Rev. Letters 25, 1622 (1970). The
    paper on Fe^{57m} has just appeared in Phys. Rev. Letters
    26, 563 (1971).

33. D. Sengupta and J. O. Artman, Phys. Rev. (to be published).

MÖSSBAUER STUDY OF POLYNUCLEAR IRON III COMPOUNDS

W. M. Reiff

Department of Chemistry
Northeastern University
Boston, Massachusetts   02115

The use of Mössbauer spectroscopy in the study of
complex polynuclear iron III systems is reviewed.
Emphasis is placed on both weak and strong exchange
coupled intramolecular antiferromagnetic systems with
iron III ions in a variety of coordination environments.
The correlation of the isomer shifts of unperturbed
Mössbauer spectra with variable temperature susceptibility
data and Near Infrared electronic spectra is shown to
conclusively eliminate the spin 1/2 ground state for the
individual iron atoms in the foregoing systems. The
failure of isomer shift and quadrupole splitting data to
unambiguously distinguish quartet and sextet ground
states and the similar draw back to susceptibility studies
are discussed. Magnetically perturbed Mössbauer spectra
are seen to be generally more useful in determining the
nature of the ground spin state and structure of polynuclear
iron systems. In this connection, the magnetic perturbation
technique is found to be a particularly simple but
powerful method for determining the presence or absence of
intramolecular ordering in iron systems that are
otherwise magnetically dilute. This is illustrated in
the non-observation of large internal hyperfine fields
in the perturbed spectra of weak ($J \sim -10$ cm^{-1}) as well as
strong ($J \sim -100$ cm^{-1}) coupled polynuclear complexes in
applied fields up to 100 kilo-Gauss. Similar spectra of
magnetically dilute paramagnetic iron salts at 4.2°K show
large augmentation of applied fields.

213

INTRODUCTION

The present article is in no way meant to be a complete review. In general the study of polynuclear complexes of transition metals is one of renewed and rapidly growing interest from both theoretical [1,2] and biological [3,4] points of view. Iron, unlike other first transition series elements shows a rather easily observed Mössbauer effect which is particularly helpful in the study of its complex polynuclear systems. The purpose of this article is to point out some of the problems as well as advantages in the use of Mossbauer techniques in the study of binuclear iron III complexes. It is also hoped to correlate some of the more important results of Mössbauer studies with those of other techniques typically used in the study of polynuclear metal complexes. The results of such studies are summarized and contrasted with particular attention to magnetic susceptibility and electronic spectra measurements.

MAGNETIC SUSCEPTIBILITY STUDIES

The initial method of study of polynuclear complexes of paramagnetic ions is usually the measurement of powder susceptibilities as a function of temperature. [5,6,7,8] Two extreme cases are possible for binuclear complexes of iron III. The first is that in which the magnetic exchange interaction between coupled metal ions is strong resulting in highly reduced susceptibilities per metal atom. Corresponding to this the effective magnetic moment per metal is usually well below the spin only expected. There are few [1] well characterized examples of polynuclear iron complexes exhibiting positive exchange interaction, i.e. intramolecular ferromagnetism and we do not consider them here. The other extreme of magnetic behavior is that in which antiferromagnetic exchange is weak such that highly reduced susceptibilities are evident only at very low temperatures. The types of behavior just discussed are summarized in Figures 1, 2, and 3 for some iron III complexes.

A number of mechanisms of exchange interaction are possible [1] in polynuclear iron complexes including direct metal-metal interaction as well as super exchange via one or more bridging ligands. It is complexes of the latter type that will be of primary concern in the present article with the bridge consisting of one or two oxygen atoms, hydroxyl

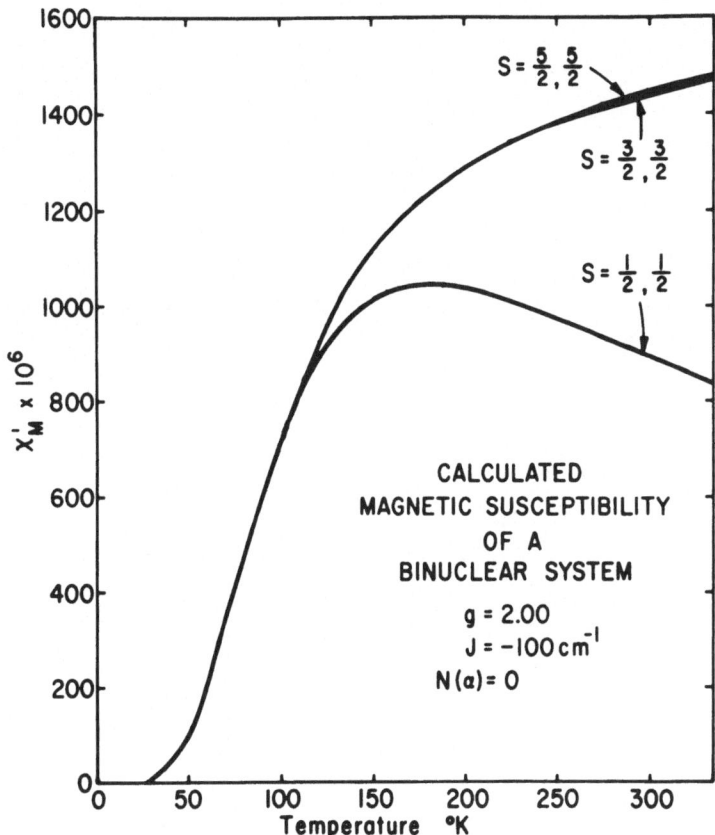

Figure 1 - Calculated molar susceptibilities for strong exchange coupled iron III centers.

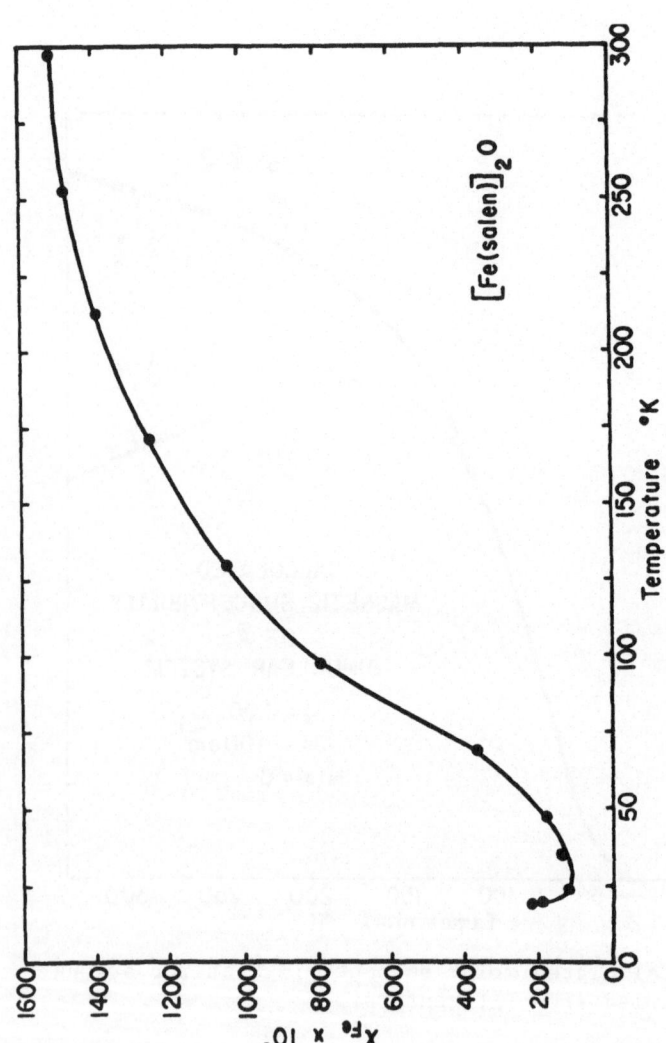

Figure 2 - Observed temperature dependence of the molar susceptibility of [Fe(salen)]$_2$O. Best fit for T > 50 °K is for J = -100 cm^{-1}, g = 2.0 Nα = 0. (Adapted from reference 8)

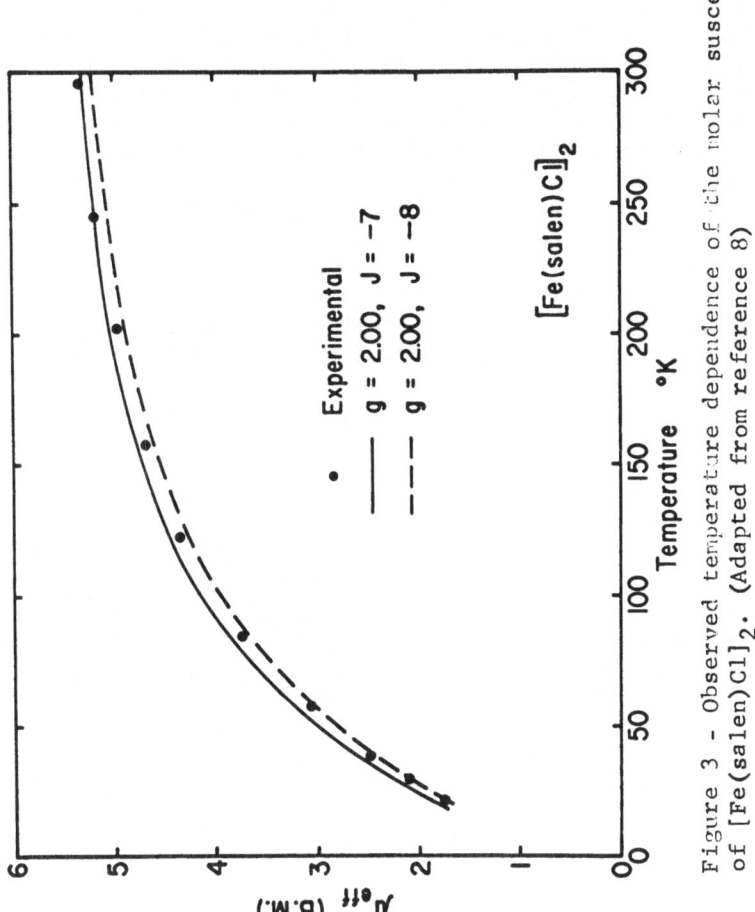

Figure 3 – Observed temperature dependence of the molar susceptibility of [Fe(salen)Cl]₂. (Adapted from reference 8)

groups or sulfur atoms. Evidence as to the nature of the
bridging usually comes from infrared [9] and/or direct single
crystal x-ray study.

Figure 1 shows the general temperature variation of
molar susceptibility for a binuclear complex in which relatively strong (J $\approx$ -100 cm^{-1}) antiferromagnetic exchange is
operative. The curves are calculated using Van Vleck's dipolar coupling model [10] for intramolecular antiferromag-
netism. In each curve both metal ions are assumed to have
the same spin and for convenience the second order Zeeman
contribution to susceptibility is taken as zero. The spin
states shown are possible for iron III, although the intermediate quartet state is somewhat rare. All three curves are
calculated for g = 2, although this value of g is usually
most appropriate to the S = 5/2 (sextet) iron III. It appears
from Figure I that powder susceptibility measurements over
the ordinary temperature range (80 to 300 °K) can adequately
distinguish S = 1/2 from S = 3/2 or 5/2 for each iron III of
a binuclear complex. On the other hand, it is also evident
that when exchange is strong the quartet and sextet states
will not diverge significantly in χ_M' until well above room
temperature at which point there would probably be decompo-
sition. The other alternative for distinguishing spin states
is to look at low temperature behavior where Nα is expected
to approach zero for the S = 5/2, 5/2, i.e. Nα is expected to
be close to zero, while the S = 3/2, 3/2 may level off [8]
to some nonzero value of Nα of the order of 200-300 c.g.s.
units. This approach is shown in Figure 2 where the best fit
to χ_M' vs T for T greater than approximately 50 °K was for
g = 2, J = -100 cm^{-1} and S = 5/2, 5/2 or 3/2, 3/2. The system considered, [Fe(salen)]$_2$O contains five coordinate iron
with each iron in the square pyramid of tetra-dentate salen^{-2}
(salen = N,N-ethylenebis-salicylideneiminate)) and bridging
oxygen with the Fe-O-Fe angle equal to 139°.[11] The structure of the ligand salen is seen in Figure 4 for the system
[Fe(salen)Cl]$_2$, another dimeric iron III complex of this ligand for which the exchange is weaker and the Fe-O-Fe angle
equal to 90°.[12,13] In Figure 2, an increase in χ_M' in the
region of 25 °K is evident. Such an increase in χ_M' is believed due to the presence of trace mononuclear paramagnetic
impurity. [8] In any event, increases in χ_M' at very low
temperatures will invalidate or at least make ambiguous
attempts to distinguish S = 3/2 from S = 5/2 spin state on
the basis of limiting values of second order Zeeman contribution to χ_M'. In spite of the preceding ambiguities in

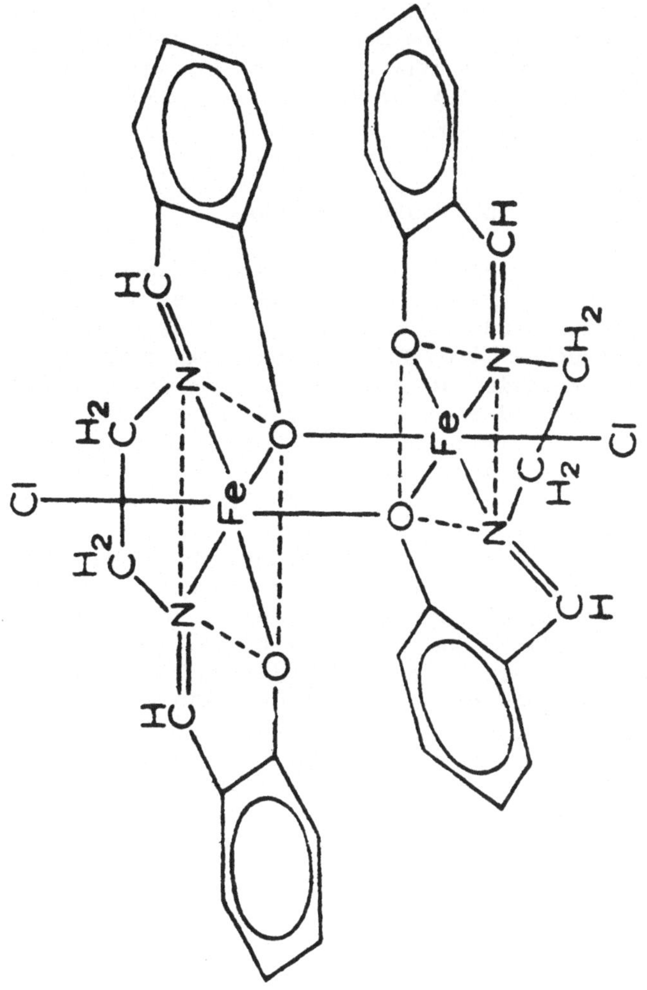

Figure 4 - Schematic representation of the dimer [Fe(salen)Cl]$_2$

susceptibility studies, the corresponding "best fit" g values close to 2 tend to favor sextet ground states for strong ex- change coupled systems such as $[Fe(salen)]_2O$ and a number of other similar systems in the literature.[8] The results of powder susceptibility investigations are more conclusive for weakly interacting iron III ions. Figure 3 shows this clear- ly for $[Fe(salen)Cl]_2$. The moment at room temperature (5.34 B.M.) is only slightly below that expected for high spin iron III (5.92 B.M.) and confirms a sextet ground state for each ion.

ELECTRONIC SPECTRA

While simple electronic spectroscopy (near infrared, visible, uv) is not generally as powerful as Mössbauer spectroscopy for study of iron III compounds, the electronic spectra of binuclear iron III complexes are atypical and de- serve brief discussion at this point. The solid state re- flectance spectra of some strong exchange coupled complexes containing the polyimine ligands phenanthroline and terpyri- dine are shown in Figure 5. [7] The absorptions above 15,000 cm^{-1} are charge transfer and ligand transitions. The well-defined absorptions at 10,000 cm^{-1} are also present in solution having extinction coefficients ≈6 and are d-d in nature. For the ligands involved in these complexes the absorptions at 10,000 cm^{-1} are too low to be reasonably assigned to doublet → doublet transitions. On the other hand, if the iron of the preceding systems is high spin (as indicated by consideration of all available types of data), then the observed extinctions are nearly an order of magni- tude too large and unexpected. This is the case since all transitions (sextet → quartet, or sextet → doublet) are nominally spin forbidden for high spin iron III. The en- hancement of spin forbidden transitions through nondilute magnetic behavior has been observed and explained for simple salts such as $KMnF_3$. [14] Antiferromagnetic coupling is be- lieved to occur through Mn-F-Mn bridging resulting in so- called "pair transitions" of higher than normal intensities. A similar phenomenon is probably operative in the foregoing binuclear iron III complexes, however on an intramolecular basis. In any event it is seen that exchange interaction in binuclear iron III complexes has a strong effect on their electronic spectra and we might expect a similar situation to obtain for their Mössbauer spectra. [8] This is the case, especially as regards the temperature dependence of line

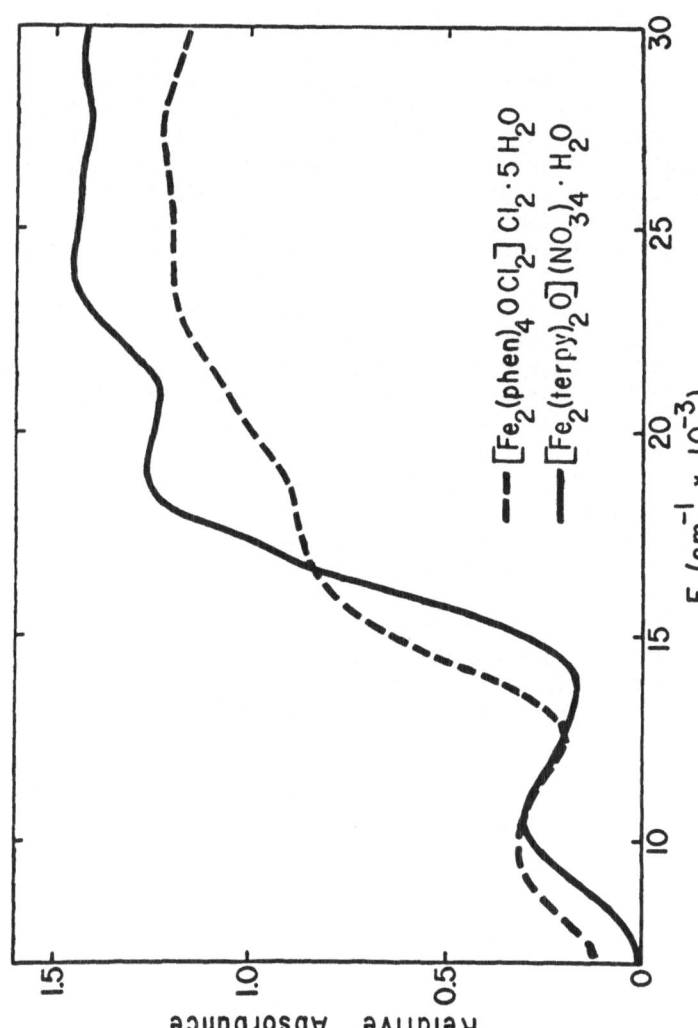

Figure 5 - Solid state, diffuse reflectance spectrum of $[Fe_2(phen)_4OCl_2]Cl_2 \cdot 5H_2O$ and $[Fe_2(terpy)_2O](NO_3)_4 \cdot H_2O$ (terpy = 2,2',2''-terpyridine). (Adapted from reference 7)

width asymmetry to be discussed subsequently. The temperature dependence of the line width asymmetry will be seen to
be directly related to the strength of exchange interaction.

UNPERTURBED MÖSSBAUER SPECTRA

The inconclusive nature of susceptibility results for
strongly coupled binuclear iron III systems has prompted this
author [7,8] and several other investigators [15,16,17] to
study such systems using unperturbed Mössbauer spectra. The
basis of such studies is the correlation of isomer shifts of
the order of 0.3 to 0.5 mm/sec (relative to iron foil) with
relatively small quadrupole splittings expected (0.5 to 0.9
mm/sec) for high spin iron III having a nominally spherically
symmetric ^6A ground term. Thus one can readily distinguish
low spin iron III from high spin by the former's near zero
values of isomer shift and as a rule, larger quadrupole effects.
This is especially the case for simple mononuclear complexes
of polyimine ligands such as bipyridine or phenanthroline.
[18] Similar isomer shift-quadrupole splitting correlations
have been made for high and low spin iron II. When one ad-
mits the possibility of intermediate spin states for iron
II or iron III, isomer shift-quadrupole splitting-spin state
correlations can become ambiguous. For example in the recently demonstrated triplet iron II compounds Fe(phenanthroline)$_2$(malonate) and corresponding oxalate, the values of
the isomer shifts and quadrupole splittings [19] are not significantly different from the tris phenanthroline iron II
compounds which have spin singlet ground states. [18]

In the case of oxo-bridged binuclear complexes, matters
are further complicated by the large spread of observed quadrupole splittings ranging from ≈ 0.7 to 1.5 mm/sec for Schiff
base systems [7,8] such as [Fe(salen)]$_2$O and [Fe(salen)Cl]$_2$
to as large as 2.4 mm/sec [7] for polyimine systems such as
[Fe(terpy)]$_2$)(NO$_3$)$_4$·H$_2$O. For the latter system the room
temperature isomer shift and quadrupole splitting are similar
to those for known quartet iron III systems such as monochlo-
robis(N,N-diethyldithiocarbamato)iron III [20] demonstrating
the danger of spin state assignments solely on the basis of
parameters of unperturbed Mössbauer spectra.

Recent studies of binuclear [21] and trinuclear [22]
hydroxo-bridged iron III systems whose magnetic data confirm
S = 5/2 iron centers generally show small quadrupole effects

expected for high spin iron III. The splittings range from
0.6 to no larger than 1.0 mm/sec. Similar parameters have
been observed [15] for the halo-bridged species such as Di-
μ-chloro-tetrachlorobisphenanthrolinedi-iron III:

$$
\begin{array}{c}
\hspace{3em}\text{Cl}\hspace{3em}\text{Cl} \\[2pt]
\left(
\begin{array}{c}
\text{N}\;|\;{}_{/}\text{Cl}\;|\;\text{N} \\
\hspace{2em}\backslash\;/\;\backslash\;/ \\
\text{Fe}\hspace{2em}\text{Fe} \\
\hspace{2em}/\;\backslash\;/\;\backslash \\
\text{N}\;|\;{}^{\text{Cl}}\;|\;\text{N} \\
\end{array}
\right)
\hspace{1em}\left(\begin{array}{c}\text{N}\\ \text{N}\end{array}\right. = \text{phenanthroline} \\[2pt]
\hspace{3em}\text{Cl}\hspace{3em}\text{Cl}
\end{array}
$$

It appears that values of the quadrupole splitting significantly greater than 1 mm/sec are limited mainly to oxo-
bridged systems where varying degrees of strong metal-oxy-
gen double bonding are possible.[23]

Some binuclear sulfur bridged species have been investigated and found to have large quadrupole effects providing
an exception to the foregoing statement. For example the
dimeric tetrachlorobenzene-1, 2-dithiolate anion of
$[n\text{-}Bu_4N]_2[Fe\{S_2C_6Cl_4\}_2]_2$ containing five coordinate iron III
with two bridging sulfur atoms has a room temperature quadrupole splitting of 3.02 mm/sec, one of the largest so far reported for five coordinate iron III. [24] Although detailed
magnetic data have not been published, the preceding and
similar systems are believed to contain S = 1/2 iron centers.
[25] Thus larger quadrupole effects are not unexpected.
The very large value in the above example may be due to nearly additive "valence" and "lattice" contributions to the total electric field gradient tensor.

LINE AREA AND WIDTH ASYMMETRY

The general shape and related temperature dependence
of unperturbed Mössbauer spectra for polynuclear iron III
systems are highly informative and should be considered at
this point. For essentially all of the systems previously
discussed or referenced, the spectral patterns and corresponding computer fits are consistent with a unique type of
iron center present. That is, a single quadrupole doublet
with varying degrees of temperature dependent line width
$(\Gamma_-/\Gamma_+ \neq 1)$ and/or line area $(A_-/A_+ \neq 1)$ asymmetry is usually
observed. The presence of a single kind of iron is also
corroborated by several single crystal x-ray studies.[11,12,
13] Considering typical coordination environments in poly-

nuclear iron III complexes, anisotropy of the recoil free
fraction (Goldanskii-Karyagin effect) and accompanying area
asymmetry are not unexpected. This area asymmetry usually
decreases significantly as the temperature is decreased to
80 °K and essentially disappears at 4.2 °K.

The temperature dependence of the line width asymmetry
can be much more complex and the following comments apply
for high spin iron III. First of all, line width asymmetry
will depend on the sign and magnitude of zero-field splitting
separating the Kramers doublets of the ground spin state.
Using ferric hemin as an example, Blume [26] has shown that
for positive zero-field splitting line width asymmetry will
be most significant at higher temperatures where the more
slowly relaxing ±3/2, and ±⁻/2 Kramers doublets are occupied
giving longer spin-spin relaxation times relative to nuclear
precession. In the case of negative zero-field splitting as
in many of the mononuclear bis-bialkyldithiocarbamate iron
III halides, [27,28] line width asymmetry increases with decreasing temperature.

For the polynuclear iron III complexes known to this
author, line width asymmetry is greatest at higher temperatures and decreases with temperature. In careful studies
of [Fe(salen)Cl]$_2$ and [Fe(salen)]$_2$O, Buckley et al. have
shown that spin-spin relaxation for antiferromagnetically
coupled complexes is intermolecular in nature. [29,30]
Further, the relaxation in such complexes is more properly
described in terms of the total spin (sum of individual ion
electron spins) and zero-field splitting of total spin states
of the dimers or trimers, etc., as opposed to a description
in terms of Kramers doublets of individual ions. In the case
of negative exchange interaction the lowest energy state is
a total spin equal to zero level. Corresponding to this
level the internal hyperfine field vanishes. Thus with decreasing temperature and progressive population of such ground
states, spin-spin relaxation broadening and line width asymmetry will disappear. This should occur regardless of the
sign of zero-field splitting of the ground electronic spin
state of the individual iron nuclei composing the polynuclear
system. This is, of course, true if there are no levels
within kT of the ground state. Increasing the temperature
results in population of levels for which the total spin is
not zero, e.g. S_{total} = 1,2,3,4,5 as in the case of exchange
coupled high spin iron III centers. The spin-spin relaxation
time between different polynuclear units increases resulting

in line width asymmetry

The other important factor to be considered in the discussion of line width asymmetry is which component $\pi(|1/2,\pm1/2\rangle \rightarrow |3/2,\pm3/2\rangle)$ or $\sigma(|1/2,\pm1/2\rangle \rightarrow |3/2,\pm1/2\rangle)$ of the quadrupole doublet undergoes initial relaxation broadening. For the simplest case, i.e. an axially symmetric electric field gradient tensor, the line which broadens first depends on the orientation of the principal component of the electric field gradient tensor (V_{zz}) relative to relaxing internal field. Blume and Tjon have shown that if the internal field (H_n) is relaxing parallel to V_{zz}, it is the π transition which undergoes initial broadening. For H_n perpendicular to V_{zz}, the σ transition broadens first. [31]

In summary, as regards the line width asymmetry of binuclear iron III compounds, two simple extremes are observed. The first is the case of strong exchange coupling ($J \approx -100$ cm^{-1}) corresponding to which little or no line width asymmetry is evident even at room temperature. In the case of weak coupling ($J \approx -10$ cm^{-1}), the line width asymmetry does not disappear until lower temperatures where there is significant depopulation of nonzero total spin levels. An example of the type spectrum observed for the latter situation is shown at the top of Figure 6. At room temperature, the line at positive velocity is broadened relative to that at negative, i.e. $\Gamma_-/\Gamma_+ < 1$. At 4.2 $^{\circ}K$ the symmetric spectrum at the top of Figure 6 results from extensive population of the $S_{total} = 0$ level in the dimeric units. As will be discussed subsequently the sign of V_{zz} in this system is negative. Hence it is the σ transition at positive velocity which is broadened at higher temperatures indicating an internal field fluctuating perpendicular to V_{zz}.

MAGNETICALLY PERTURBED SPECTRA

The magnetically perturbed spectra of one of the binuclear systems previously discussed, $[Fe(salen)Cl]_2$, are given in the center and bottom of Figure 6. The spectra were determined at 4.2 $^{\circ}K$ in longitudinal fields of 9 and 26 kilo-Gauss respectively. It is evident that V_{zz} is negative, with a well defined, narrow lined triplet at positive velocity and a more broadened doublet with slight center pattern intensity at negative velocity. [32] The general shape of the spectrum at 26 kilo-Gauss is consistent with a small non-zero asymmetry

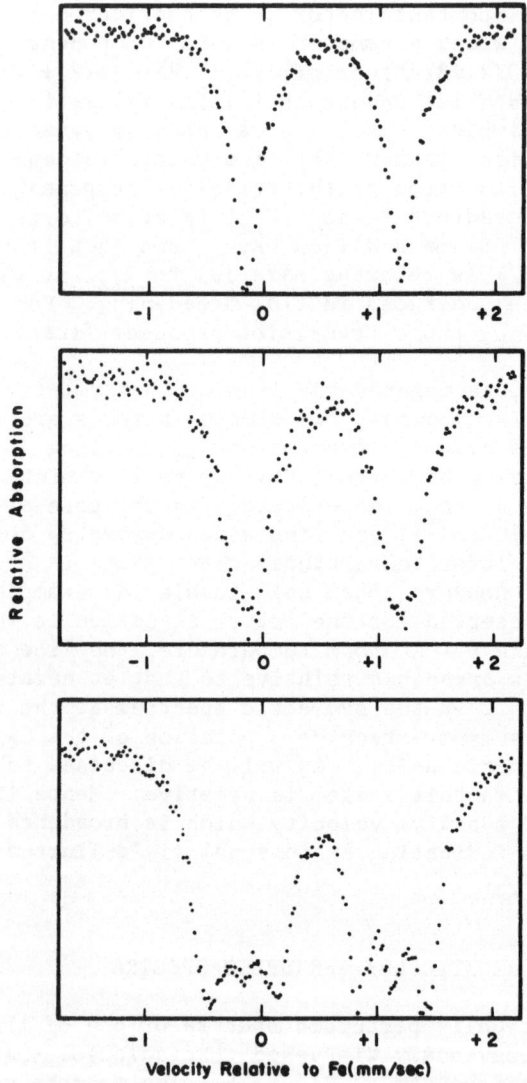

Figure 6 - Top - Mössbauer Spectrum of [Fe(salen)Cl]$_2$ at
4.2 ^{o}K, H$_{applied}$ = 0.   Center - H$_{applied}$ = 9 kilo-Gauss.
Bottom - H$_{applied}$ = 26 kilo-Gauss.

parameter ($\eta \approx 0.2$ to 0.3).

A feature of the magnetically perturbed spectra that is of special importance as regards the general behavior of typical antiferromagnetically coupled polynuclear systems should be discussed now. It is observed that the internal hyperfine field as calculated from the triplet splitting (bottom of Figure 6) is nearly equal to the applied field with an experimental error. This is not ordinarily the case for the perturbed spectra of simple mononuclear paramagnetic systems. In the latter, large internal fields are observed corresponding to small applied fields at low temperatures. Since writing this review, the author has learned that Buckley and co-workers have also determined magnetically perturbed Mössbauer spectra for the dimer $[Fe(salen)Cl]_2$ and at even larger applied fields (50 and 90 kilo-Gauss) with essentially the same results as reported here. [33] That is, very large applied fields do little to perturb the antiferromagnetic coupling for even weak exchange coupled systems such as $[Fe(salen)Cl]_2$. A similar and not unexpected result has been observed for the perturbed spectra of strongly coupled systems such as $\{[Fe(phen)_2Cl]_2O\}Cl_2$. [23] Thus it is seen that the internal field of perturbed spectra is diagnostic of intramolecular ordering as in the present systems. This is the case provided one has other data ruling out the presence of diamagnetic low spin iron II.

Buckley et. al. have determined the magnetically perturbed Mössbauer spectra for the Schiff base iron III adduct $Fe(salen)Cl \cdot 1/2 \, CH_3NO_2$ obtained by slow crystallization of the dimer $[Fe(salen)Cl]_2$ from CH_3NO_2. [34] Other investigators have assumed such adducts to be monomeric on the basis of room temperature susceptibility, infrared and unperturbed Mössbauer spectra. [16] The perturbed spectrum of Fe(salen) $Cl \cdot 1/2 \, CH_3NO_2$ obtained by slow crystallization shows no large internal field ($H_{applied} \approx H_{internal}$) indicating that the material is still dimeric in agreement with variable temperature susceptibility data.

This author has determined the perturbed Mössbauer spectra of the materials obtained by rapid crystallization of $[Fe(salen)Cl]_2$ from various solvents such as C_5H_5N and CH_3NO_2. [35,36] The spectra are typical of those expected of a material which is primarily monomeric. That is, they are broadened and show large augmentation of the applied field. Further, the sign of V_{zz} is now positive and $\eta \approx 0$ as might be

expected for true monomers with the formulation Fe(salen)Cl·
X C_5H_5N or CH_3NO_2 and neither solvent coordinated. In summary,
it appears that the techniques of perturbed spectra can quite
nicely distinguish varying degrees of intramolecular order.

CONCLUSIONS AND SOME SUGGESTIONS FOR FUTURE WORK

Problems remaining and related to the Mössbauer study
of polynuclear iron complexes are numerous. The technique
of Mössbauer spectroscopy is one of the more useful methods
of looking at such systems, although it is important that
its results be carefully correlated with those of other methods.

The Mössbauer effect has been valuable in the study of
complex polynuclear heme - iron III proteins. [37] In some
systems unexpectedly sharp (small line width asymmetry and
narrow lines) spectra are observed at room temperature. This
corresponds to strong antiferromagnetic exchange as discussed previously in this article. The construction of bi-
and trinuclear complexes as models for more complicated natural polymeric systems, e.g. the non-heme iron-sulfur proteins
is certainly an area where the Mössbauer effect will be in-
dispensable as a characterization tool. In an attempt to
more fully understand the relaxation phenomenon taking place
in polynuclear systems, room temperature magnetically perturbed spectra might be helpful. Finally, the evident lack
of single crystal studies point to this area as one of possible important future research in the Mössbauer study of
polynuclear systems.

ACKNOWLEDGMENTS

The author wishes to thank Dr. M. L. Good and R. L.
Prados for a preprint of some of their recent experimental
results. He also wishes to thank R. L. Collins for many
fruitful discussions during his stay at the University of
Texas.

BIBLIOGRAPHY

1. E. Sinn, Coordin. Chem. Rev., $\underline{5}$ 313 (1970).
2. P. W. Ball, Coordin. Chem. Rev., $\underline{4}$ 361 (1969).
3. T. G. Spiro and P. Saltman, "Structure and Bonding" $\underline{6}$, 116 (1969).
4. J.C. M. Tsibris and R. W. Woody, Coordin. Chem. Rev. $\underline{5}$, 417 (1970).
5. J. Lewis, F. E. Mabbs, and A. E. Richards, J. Chem. Soc. A., 1014 (1967).
6. A. V. Khedekar, J. Lewis, F. E. Mabbs, and H. Weigold, ibid., 1561 (1967).
7. W. M. Reiff, W. A. Baker, Jr., and N. E. Erickson, J. AM. Chem. Soc., $\underline{90}$, 4794 (1968).
8. W. M. Reiff, G. J. Long, and W. A. Baker, Jr., ibid., $\underline{90}$, 6347 (1968).
9. F. A. Cotton and R. M. Wing, Inorg. Chem., $\underline{4}$ 867 (1965).
10. J. H. VanVleck, "Electric and Magnetic Susceptitibilities" Oxford Univ. Press.
11. M. Gerloch, E. D. McKenzie and A. D. C. Towl, Nature, $\underline{220}$, 906 (1968).
12. M. Gerloch, J. Lewis, F. E. Mabbs, and A. Richards, Nature $\underline{212}$, 809 (1966).
13. M. Gerloch and F. E. Mabbs, J. Chem. Soc. $\underline{A}$, 1900 (1967).
14. J. Ferguson, H. J. Guggenheim, and Y. Tanabe, J. Phys. Soc. Japan, $\underline{21}$, 692 (1966).
15. R. R. Berrett, B. W. Fitzsimmons and A. A. Owusu, J. Chem. Soc. A., 1575 (1968).
16. G. M. Bancroft, A. G. Maddock, and R. P. Randl, ibid., 2939 (1968).
17. J. A. Bertrand, J. L. Breece, A. R. Kalyanaman, G. J. Long, W. A. Baker, Jr., J. Am. Chem. Soc. $\underline{92}$ 5233 (1970).
18. R. L. Collins, R. Pettit, and W. A. Baker, Jr., J. Inorg. Nucl. Chem., $\underline{28}$, 1001 (1966).
19. E. Konig and K. Madega, J. Am. Chem. Soc., $\underline{88}$, 4528 (1966).
20. A. Hudson and H. J. Whitfield, Inorg. Chem., $\underline{6}$ 712 (1967).
21. H. J. Shugar, G. R. Rossman, H. B. Gray., J. Am. Chem. Soc., $\underline{91}$, 4564 (1969).
22. R. Prados and M. L. Good, J. Inorg. Nucl. Chem. (1971) submitted, and private communication.
23. W. M. Reiff, J. Chem. Phys. (1971) in press.
24. T. Birchall and N. N. Greenwood, J. Chem. Soc. $\underline{A}$, 286 (1969).

25. J. A. McCleverty, J. Locke, and N. Connelly, Nature,
     216, 999 (1967).
26. M. Blume, Phys. Rev. Letters, 18, 305 (1967).
27. H. H. Wickman and C. F. Wagner, J. Chem. Phys., 51,
     435 (1969).
28. R. L Ake and G. M. Loew, J. Chem. Phys. 52, 1098 (1970).
29. A. N. Buckley, G. V. H Wilson, and K. S. Murray,
     Solid State Comm., 7, 471 (1969).
30. A. N. Buckley, G. V. H. Wilson, and K. S. Murray, Chem.
     Comm., 718 (1969).
31. M. Blume and J. A. Tjon, Phys. Rev., 165, 446 (1968).
32. R. L. Collins and J. C. Travis, "Mossbauer Effect
     Methodology", 3, 123 (1967).
33. A. N. Buckley, I. R. Herbert, B. D. Rumbold, G. V. H.
     Wilson and K. S. Murray, J. Phys. Chem. Solids, 31,
     1423 (1970).
34. A. N. Buckley, B. D. Rumbold, G. V. H. Wilson and K.
     S. Murray, J. Chem. Soc., A, 2298 (1970).
35. W. M. Reiff, Abstracts Physical Division, Paper No. 127,
     February 1970, A.C.S. Meeting, Houston, Texas.
36. W. M. Reiff, to be submitted for publication.
37. I. W. Cohen., J. Am. Chem. Soc., 91, 1980 (1969).

METHODOLOGY

PROPERTIES OF $Ni_{21}Sn_2B_6$ AS SOURCE FOR Sn^{119} AND Sb^{121}
MÖSSBAUER EXPERIMENTS*

L. H. Bowen, K. A. Taylor, H. Z. Dokuzoguz†,
and H. H. Stadelmaier

Departments of Chemistry and Engineering
Research, North Carolina State University
Raleigh, North Carolina 27607

Because the 37 kev Mössbauer level of Sb^{121} is produced
by beta decay from Sn^{121}, sources for Sb^{121} and Sn^{119}
Mössbauer spectroscopy have much in common. Sources of
$Sn^{121}O_2$ and $CaSn^{121}O_3$ have been used primarily for Sb^{121}
spectroscopy (1,2,3) because, just as for Sn^{119}, these sources
have large recoilless fractions at liquid nitrogen temperature
and below. Due to the higher energy of the transition of
Sb^{121}, the recoilless fraction at room temperature is lower
than in the corresponding Sn^{119m} source. The recoilless
fraction and linewidth for $Sn^{121}O_2$ and $CaSn^{121}O_3$ are apparently almost the same (2), even though $Sn^{119m}O_2$ gives a
somewhat broadened line due to unresolved quadrupole splitting.

Differences in the Sb^{121} spectra from those with Sn^{119}
include the eight overlapping lines in the quadrupole interaction (a 7/2 to 5/2 transition) and the larger change in
nuclear radius, of opposite sign. Thus, the isomer shift
range for Sb^{121} is about 6 times larger than that for Sn^{119}.
Since the oxide sources give off a gamma ray characteristic
of Sb(V), absorption from a typical Sb(III) compound occurs
at -10 to -16 mm/sec, necessitating a large velocity scan

*Research supported in part by the National Science
Foundation.

†Present Address: Babcock and Wilcox Research Center,
Lynchburg, Virginia.

233

and often causing difficulties in velocity calibration. In principle, a metallic source should produce a range of isomer shifts for Sb(III) and Sb(V) more symmetrically displaced from zero velocity. However, experiments with β–Sn and Mg_2Sn have not produced useful sources (2).

Considering its favorable properties as a source for Sn^{119} experiments (4), we have prepared and tested a source of Pd_3Sn^{121} (5). The linewidths were as narrow as those from $Sn^{121}O_2$ and the expected change of about 8 mm/sec was observed in isomer shifts. However, the reduction in counting rate due to scattering by Pd and the low recoilless fraction, estimated as only about 1/3 that of the $Sn^{121}O_2$, indicated it was not suitable as an alternate source material.

For the past 10 years studies have been made on the structure and properties of *tau* phase borides with the $Cr_{23}C_6$ structure (6). The properties of these compounds are such that one would predict favorable source characteristics, in particular, for the nickel compound with tin (7). Having an ideal composition of $Ni_{21}Sn_2B_6$ ($Ni_{71.4}Sn_{6.9}B_{20.7}$), with four formula units in the face-centered cubic unit cell, the compound contains Sn at the corners of a cube whose edge is half the unit cell edge in length (position *c* in Figure 1).

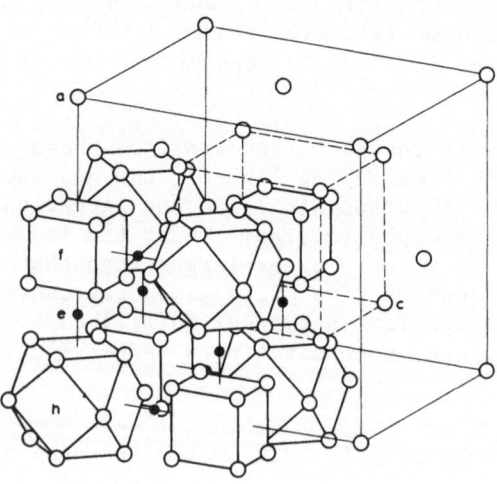

Figure 1.   Structure of *tau* phase boride.

Unlike the carbides, borides cannot have this structure
without a stabilizing species such as Sn or Sb in the
special lattice position c (6). The Sn is coordinated with
16 Ni, 4 being slightly closer in tetrahedral coordination
from the Ni cubes (position f) and 12 from the tetrahedrally
oriented cubo-octahedrons (position h). Thus, one would
expect this material to produce a single unsplit line from
a Mössbauer nucleus in position c. As $Ni_{21}Sb_2B_6$ is known
to have the same structure (8), beta decay of Sn^{121} should
have minimal effect on the Mössbauer emission from the Sb^{121}
formed. Moreover, the X-ray temperature factor is low for
Sn in this compound and the Sn radius appears compressed (6).
One would therefore predict a large recoilless fraction.
Mössbauer resonance from absorbers of $Ni_{21}Sn_2B_6$ and
$Ni_{21}Sb_2B_6$ has been reported as part of a general study of
nickel-containing alloys (5,9). The narrow linewidth and
large percent absorption from these compounds led to the
present study of sources both of $Ni_{21}Sn_2^{121}B_6$ and
$Ni_{21}Sn_2^{119m}B_6$.

EXPERIMENTAL

For making absorbers, materials with the purities 99.8%
(Ni), 99.99%(Sn), 99.98%(Sb), and 99.4%(B) were used.
Because elemental B tends to float on Ni, it is preferable
to make Ni_2B first and add the appropriate weight of that
compound, rather than use B itself. An induction furnace
was used to allow rapid heating of the samples to the final
temperature of about 1400°C in 10-15 minutes. The furnace
was turned down immediately after reaching the final temperature and the samples furnace-cooled. The samples were all
sealed in fused silica capsules under vacuum of 10^{-4} torr.
The Ni_2B was heated to 1350°C and quenched from 800° into
water. The tau phase nickel borides were heated to 1400°,
but were not quenched. Small departures from the ideal
stoichiometric composition of 6.9 atomic percent have been
reported (6), and at 800°C the range is for Sb 6.5 to 7.0%
and for Sn 6.0 to 6.8%. Weights of Ni, Ni_2B and Sn or Sb
were chosen to make single phase alloys of composition
$Ni_{72.4}Sb_{6.9}B_{20.7}$ or $Ni_{71.4}Sn_{6.6}B_{22.0}$. On the boron-rich
side of the phase diagram, any second phase formed should
not contain Sn or Sb (7,8). The material formed is brittle
and easily crushed to a fine powder (-325 mesh screen).
Examination of a number of preparations by metallographic
techniques and X-ray powder diffraction gave no evidence of

any appreciable second phase. The powder was mixed with
dextrose or polyethylene powder in a Lucite holder of uniform thickness. For some of the experiments we used as
absorbers $BaSnO_3$, made as described by Sano and Herber (10).
Commercial InSb powder (Alfa Inorganics) was also used for
absorbers. Purity of these compounds was checked by X-ray
diffraction.

For sources we used commercially-supplied Co^{57} (Pd),
$BaSn^{119m}O_3$, and Sn^{119m} (New England Nuclear Corp.). The
tau phase borides were made in a similar fashion to the
absorbers above, but could not be checked for phase purity
other than to note visually that they formed a homogeneous
boule with no weight loss. The Sn^{119m} could be used directly as the metal in making the boride. The Sn^{121} was elec-
troplated from an HCl solution onto a Ni foil. Both alloy
sources were crushed to a powder, mixed with dextrose, and
encased in Lucite. Because a mesh was not used, some non-
uniformity of particle size was present. The $Sn^{121}O_2$ source
was similar to that described earlier (1), but was annealed
for a longer period (24 hr at 1050°C) and gave almost 2
times the percent absorption of the earlier source. Each
source of Sn^{121} contained about 0.1 mCi in 100 mg Sn, and
the $Ni_{21}Sn_2^{119m}B_6$ source was about 0.3 mCi in 14 mg Sn.
The $BaSn^{119m}O_3$ had about the same activity as the latter.

Mössbauer spectra were taken with an Austin Science
Associates drive, Nuclear Data 512 channel analyzer, and
$Xe-CH_4$ proportional tube, counting the escape peak of the
Sb^{121}. For the most of the Sn^{119} runs, a $Xe-N_2$ counter
was used. Low temperature runs were made with both source
and absorber cooled to about 80°K. The data were computer-
fitted to single Lorentzian curves. The velocity scale was
calibrated using the magnetic splitting of metallic iron.

RESULTS AND DISCUSSION

The expected result of a shift towards more centrally
displaced velocity was observed with both sources. For example, the spectrum of Sb_2O_4 is shown in Figure 2. This
compound has been previously reported with $Sn^{121}O_2$ as a
source (1). A velocity scale almost twice as large was
previously required to obtain the very negative peak of the
Sb(III) absorption. Both the $Ni_{21}Sn_2^{119m}B_6$ with $Ni_{21}Sn_2B_6$

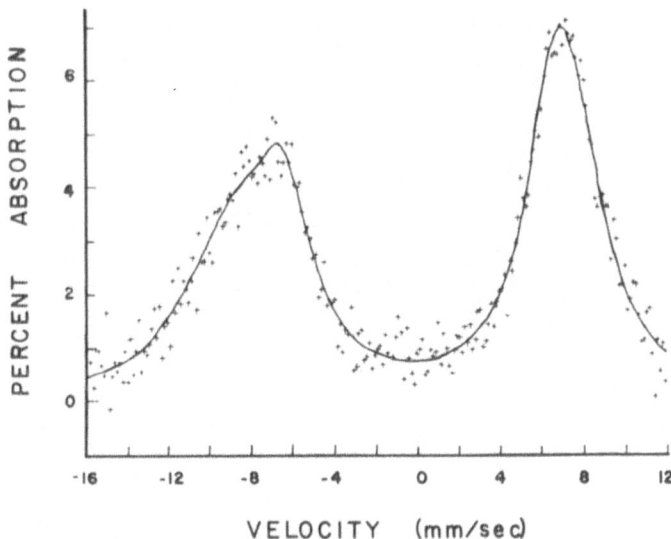

VELOCITY (mm/sec)

Figure 2.   Mössbauer absorption spectrum for Sb$_2$O$_4$ with Ni$_{21}$Sn$_2$^{121}B$_6$ source. Quadrupole splitting was assumed as a parameter, but the widths of all single peaks were constrained equal. The ratio of intensities of the Sb(III) to Sb(V) peak was 0.91 (should be 1.00 theoretically). Quadrupole splitting of the Sb(V) peak at positive velocity was small (e^{2}qQ = -6 mm/sec); in the Sb(III) peak at negative velocity it was appreciable (e^{2}qQ = +17 mm/sec). The symmetry of the Sb(III) and Sb(V) peaks relative to the alloy source is evident [δ = +7.1 mm/sec for Sb(V) and -7.9 for Sb(III)].

as absorber and Ni$_{21}$Sn$_2$^{121}B$_6$ with Ni$_{21}$Sb$_2$B$_6$ as absorber gave single symmetric peaks essentially at zero velocity (Figure 3). Due to the large natural linewidth of Sb^{121} (2.1 mm/sec), that peak is considerably broader than the one for Sn^{119}.

In order to evaluate the properties of the sources as well as the alloy absorbers, comparisons were made with several standard compounds. In addition, a series of runs varying thickness of the absorber were made with each alloy and with BaSnO$_3$. Two absolute methods were used for calculating recoilless fractions (f values). The first method is the variation of linewidth with thickness:

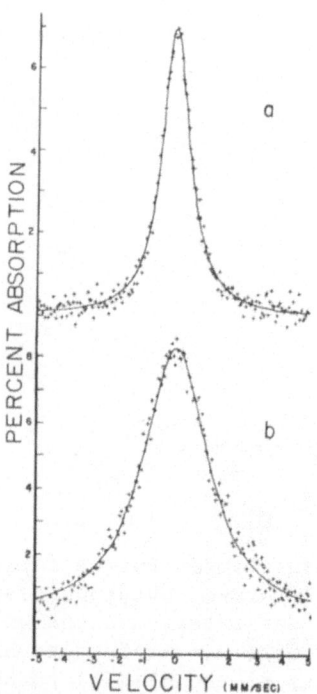

Figure 3.   Mössbauer absorption spectra for (a) $Ni_{21}Sn_2{}^{119m}B_6$
source with absorber of the same alloy.
(b) $Ni_{21}Sn_2{}^{121}B_6$ source with absorber of
$Ni_{21}Sb_2B_6$. Both spectra were fitted with single
Lorentzian curves with isomer shift at approximately zero velocity.

$\Gamma(\exp) = \Gamma_0 + 0.27 \, \Gamma_\gamma t$, where the effective thickness,
$t = f_a N_a \sigma_0 x$ (11). However, due to the difficulty of re-
produceability of linewidths and the small variation over
the thickness of absorbers used (2 to 8 mg/cm^2 either in
Sn or Sb), this method proved imprecise. The other method,
the area method (12), uses the equation $A = \pi/2 \, \Gamma_\gamma f_s L(t)$,
where $L(t) = t$ for a thin absorber. Since the area under a
Lorentzian absorption curve is $A = \pi/2 \, \varepsilon\Gamma(\exp.)$, where ε is
the fractional effect magnitude, this equation can be used
to calculate the product $f_s f_a$, and extrapolation of the
values to $t = 0$ should give the true $f_s f_a$. Of course, one
must convert observed percent effects into ε by correcting
for background. For the Sb^{121} this is straightforward, as
the background under the escape peak is rather uniform.

For Sn^{119}, a correction must be made for the Sn X-rays
which are also counted. This correction was obtained by
using a set of Pd filters of varying thickness (made from
palladium dimethylglyoxime), and observing the reduction in
counting rate as a function of Pd thickness. The calculated
ratio $\gamma/(\gamma + X$-ray) in the peak should be precise if extrapolated to infinite thickness. In addition to the absolute
methods above, the ratio of areas obtained with a common
source or absorber could be used in some cases to obtain
f values.

Sn^{119} Results

Table I gives a comparison of these at liquid nitrogen
temperature. There is only a slight improvement in percent
effect and linewidth in going from β—Sn to the alloy, either
source or absorber. However, without reducing the percent
effect the alloy reduces the linewidth over $BaSnO_3$ by about
0.1 mm/sec for comparable thicknesses. The δ values show

Table I. Comparison of Sn^{119} spectra at 80°K with different
source-absorber combinations.

Source	Absorber	Absorber Thickness (mg Sn/cm²)	% Effect	Γ (±0.01 mm/sec)	δ (±0.005 mm/sec)
Sn	Sn	8.7	9.3	1.14	0.032
$ST^{(a)}$	Sn	8.7	8.5	1.25	2.533
$AY^{(a)}$	Sn	8.7	10.0	1.12	1.057
Sn	AY	7.1	8.1	1.17	-1.044
ST	AY	7.1	9.5	1.26	1.506
AY	AY	7.1	9.8	1.08	-0.019
AY	ST	7.1	8.8	1.24	-1.519

[a] ST stands for $BaSnO_3$, AY for $Ni_{21}Sn_2B_6$.

the expected trends, and the alloy is characteristic of a
metallic Sn species with rather low s electron density (9).
A much more drastic difference in percent effect occurs at
room temperature, where the combination Sn source-alloy absorber gave only 0.7%, while the alloy source with the same
absorber gave 3.2%. Since $BaSnO_3$ gave even higher percent
effects, the alloy source is not favored for use at room
temperature.

Calculated f values by the methods indicated are shown
in Table II. Our assumption has been that $f_s = f_a$ for the

Table II. Recoilless fraction for $Ni_{21}Sn_2{}^{119}B_6$ calculated
by different methods.

Source	Absorber	Method	f(absorber)	f(source)
T = 80°K				
AY	AY	Line Width	0.6 ± 0.1	
AY	AY	Area	0.58 ± 0.05	0.58 ± 0.05
AY	ST	Area	(a)	0.55 ± 0.05
T = 295°K				
AY	AY	Area	0.34 ± 0.05	0.34 ± 0.05
ST	AY	Relative Area	0.36 ± 0.05	(a)
AY	ST	Area	(a)	0.37 ± 0.05

(a)Values for $f(BaSnO_3)$ taken from Reference 10: 0.65 at
80°K and 0.58 at 295°K.

alloy pair. The agreement in the area method using both
$BaSnO_3$ and the alloy absorbers is good. Results at room
temperature are also consistent. It appears the ratio
$f(alloy)/f(BaSnO_3)$ = 0.9 at 80°K and 0.6 at 295°K. The linewidth method is in agreement with these numbers. The extrapolated value of Γ_0 for the alloy pair was disappointingly
high at 0.91 ± 0.05 mm/sec. With $BaSnO_3$ absorbers, a high
value of Γ_0 = 0.95 mm/sec was also obtained. However, in
order to cool both source and absorber, we used a vertical

system with the source at the end of a 6" rod extending into the cryostat. To test the broadening effect due to vibrational noise, we ran a series of room temperature measurements in horizontal geometry with the source mounted directly on the drive. The extrapolated linewidth obtained for the alloy pair was $\Gamma_0 = 0.71$ mm/sec. Sano and Herber (10) report for $BaSnO_3$ $\Gamma_a = 0.33$ and $\Gamma_s = 0.37$ mm/sec, giving $\Gamma_0 = 0.70$ mm/sec for a $BaSnO_3$ source-absorber pair. It is apparent that the inherent linewidth of the alloy is at least as narrow as this value.

Sb^{121} Spectra

In Table III the alloy source is compared with our $Sn^{121}O_2$ source. The percent effects are somewhat greater, although

Table III.   Comparison of Sb^{121} spectra at 80°K with different source-absorber combinations.

Source	Absorber	Absorber thickness (mg Sb/cm^2	% Effect	Γ (±0.05 mm/sec)	δ (±0.05 mm/sec)
AY[a]	InSb	10.8	5.6	2.98	-1.74
SnO_2	InSb	12.9	4.4	2.85	-8.54
AY	AY[a]	5.1	8.3	2.86	+0.03
SnO_2	AY	7.5	8.1	2.98	-6.95

[a]AY stands for $Ni_{21}Sb_2B_6$ as absorber, or $Ni_{21}Sn_2B_6(Sb^{121})$ as source.

the combination of narrower linewidth and a slightly smaller background correction tend to equalize the areas. The advantage of using the alloy source in studies of metallic species is that a velocity scale comparable to that for Sn^{119} can be used. As was shown in Figure 2, the velocity scale required for ionic compounds is larger, but still about a factor of 2 less than if $Sn^{121}O_2$ is used with Sb(III) absorbers.

The recoilless fractions have been calculated in a similar manner to the previous ones, and are shown in Table IV. Scatter in the linewidths prevented any conclusions to

Table IV. Recoilless fraction for $Ni_{21}Sn_2{}^{121}B_6(Sb^{121})$ and $Ni_{21}Sb_2B_6$ calculated by different methods.

Source	Absorber	Method	f(absorber)	f(source)
T = 80°K				
AY	AY	Line Width	0.5 ± 0.3	
AY	AY	Area	0.29 ± 0.05	0.29 ± 0.05
SnO_2	AY	Relative Area	(a)	0.32 ± 0.05
T = 295°K				
AY	AY	Relative Area	0.07 ± 0.02	0.07 ± 0.02
SnO_2	AY	Relative Area	(a)	0.22 ± 0.02

(a)The value of f(absorber) was assumed to calculate f(source): f(AY) = 0.29 at 80°K and 0.07 at 295°K.

be drawn from the first method. The value of $\Gamma_0 = 2.6 \pm 0.1$ mm/sec is as narrow as has been reported for Sb^{121} sources. Line broadening from the experimental arrangement likely has affected this value as well as the one for Sn^{119}. The assumption that $f_s = f_a$ is perhaps not *a priori* acceptable, since these are different compounds. However, the local environment about Sb is identical in the two alloys, and screening by Ni should make the substitution of Sn for Sb a minor effect. The fact that a similar f value for the $Sn^{121}O_2$ source was obtained is reasonable. However, this number is lower than the reported one for $CaSn^{121}O_3$ of about 0.6 (2), whereas that source and $Sn^{121}O_2$ have been reported as almost identical. A check on our calculations can be made by comparing f values for the alloy with Sb^{121} and with Sn^{119}. As shown below, this comparison is consistent with the calculated f values. Our values give f (alloy)/f $(Sn^{121}O_2)$ = 0.9 at 80°K and 0.3 at 295°K.

GENERAL CONCLUSIONS

The $Ni_{21}Sn_2B_6$ is comparable at 80°K to other sources for Sn^{119} and Sb^{121} Mössbauer spectroscopy. It has some advantage for the latter in terms of velocity range. Table V summarizes the isomer shift data. At room temperature the comparison with standard oxide sources is somewhat unfavorable. Even here, absorbers of $Ni_{21}Sb_2B_6$ or $Ni_{21}Sb_2B_6$ should be considered as possibilities for standards. These alloys are known to be quite stable. Results from different preparations in the present work were identical within experimental error. No change was seen in spectral quality over periods greater than a year. In addition, a sample of $Ni_{21}Sn_2B_6$ immersed in water for a day, heated to boiling, and evaporated to dryness, gave identical linewidth to the original material.

As a concluding note, it is interesting to compare the alloy f values by means of the equation $\ln f = -[x^2]/\bar{x}^2$. If the reasonable assumption is made that $[x^2]$ is the same for both Sn and Sb in the tau phase boride, the ratio $\ln f(Sb)/\ln f(Sn)$ would be equal to the ratio of the respective recoil energies, 2.38. Calculated ratios are at 80°K, $\ln(.29)/\ln(.58) = 2.3$, and at 295°K, $\ln(.07/\ln(.34) = 2.5$. Although

Table V. Isomer shift conversions. Average value of the shift in mm/sec between the alloy source and the absorber listed at 80°K.

Source	Absorber	Shift
Sn^{119}:	$BaSnO_3$	-1.500 ± 0.005
	$Ni_{21}Sn_2B_6$	-0.003
	β—Sn	$+1.050$
Sb^{121}:	$SnO_2(Sb^{121})$	$+6.91 \pm 0.05$
	$Ni_{21}Sb_2B_6$	-0.03
	InSb	-1.70
	Sb	-4.72

$[x^2]$ should be only approximately equal for SnO_2 and $BaSnO_3$, the ratio for these at 80° is $\ln(.32)/\ln(.65) = 2.6$. These calculations show that it is unlikely for a source for Sb^{121} to have an f value much greater than the ones reported here. The alloy sources have, as predicted, less variation in f value with temperature than a Debye solid, even though greater than that for $BaSnO_3$.

ACKNOWLEDGMENTS

The assistance of R. G. Snipes, L. E. Shaw, and R. A. Williams in chemical preparations is gratefully acknowledged. Miss. M.-L. Fiedler provided valuable help in X-ray analysis and metallurgical procedures. We appreciate the assistance of C. W. Seidel of New England Nuclear Corp. in obtaining the sample of Sn^{119m}.

REFERENCES

1. J. G. Stevens and L. H. Bowen, "Mössbauer Effect Methodo- logy", Vol. 5, I. J. Gruverman, Ed., p. 27, Plenum Press, New York (1969).
2. S. L. Ruby, "Mössbauer Effect Methodology", Vol. 3, I. J. Gruverman, Ed., p. 203, Plenum Press, New York (1967).
3. G. G. Long, J. G. Stevens, R. J. Tullbane, and L. H. Bowen, *J. Amer. Chem. Soc.*, **92**, 4230 (1970).
4. D. K. Snedicker, "Mössbauer Effect Methodology", Vol. 2, I. J. Gruverman, Ed., p. 161, Plenum Press, New York (1966).
5. H. Z. Dokuzoguz, "Sn^{119} and Sb^{121} Mössbauer Effect in Some Alloys and Intermetallic Compounds", Ph.D. Thesis, North Carolina State University, Raleigh (1969).
6. H. H. Stadelmaier, "Developments in the Structural Chemistry of Alloy Phases", B. C. Giessen, Ed., p. 141, Plenum Press, New York (1969).
7. H. H. Stadelmaier and L. T. Jordan, *Z. Metallkde.*, **53**, 719 (1962).
8. G. Hofer and H. H. Stadelmaier, *Metall*, **18**, 963 (1964).
9. H. Z. Dokuzoguz, H. H. Stadelmaier, and L. H. Bowen, *J. Less Common Met.*, **23**, 245 (1971).
10. H. Sano and R. H. Herber, *J. Inorg. Nucl. Chem.*, **30**, 409 (1968).
11. D. A. O'Connor, *Nucl. Instr. Meth.*, **21**, 318 (1963).
12. G. Lang, *Nucl. Instr. Meth.*, **24**, 425 (1963).

A TECHNIQUE FOR THE REMOVAL OF THE "BLACKNESS" DISTORTION OF MÖSSBAUER SPECTRA*

M. Celia Dibar Ure and P. A. Flinn

Carnegie Mellon University

Pittsburgh, Pennsylvania

The interpretation of Mössbauer spectra is frequently complicated by the "blackness" or "thickness" effect; that is, the fact that the observed spectrum is expressible as a sum of peaks of Lorentz shape only in the limit of a "thin" sample. Increasing the sample thickness broadens and distorts the peaks. Various procedures to correct for this effect have been given (1,2,3), but they require assumptions about the source and absorber characteristics. Our procedure requires no assumptions about either source or absorber characteristics, and provides for the removal of effects due to both sample thickness and non-ideal source peakshape. The final result of the calculation is a spectrum which would be obtained from an ideally thin sample and an ideally monochromatic source.

A spectrum in the absence of polarization effects is given by the general expression

$$(1) \quad n_e(v) = f n_0 \int_{E_1}^{E_2} S(E-v) \exp\left[-\sigma(E)t\right] dE + (1-f)n_0 + n_a$$

*This research was supported, in part, by the United States Atomic Energy Comission.

245

where E is the energy, expressed in velocity units, and v is the velocity; $n_e(v)$ is the number of counts observed at each velocity, f is the resonant fraction of the source, $S(E)$ is the normalized source shape, and E_1 to E_2 is the energy range accepted by the detector (effectively $-\infty$ to $+\infty$ compared to the line width of the peaks), n_0 is the number of gamma rays of the Mössbauer transition which would be counted at each velocity in the absence of Mössbauer absorption, and n_a is the number of additional counts (due to other radiation) which are observed. Thus there are three contributions to the number of counts in the portions of the observed spectrum free from resonant absorption (the "background" region). These are: fn_0, the number of gamma rays which were emitted without recoil, $(1-f)n_0$, the number of gamma rays (from the Mössbauer transition) emitted with recoil, and n_a, defined above. The sum of these components, $n_a + n_0$, we will call the total background, n_b. In the case of a Co^{57} in Cr source, the main contribution to n_a comes from 122 Kev gamma rays.

Factoring out $f.n_0$ and using the fact that the source shape is normalized, we may write eq. (1) as

$$(2) \quad n_e(v) = fn_0 \int_{-\infty}^{+\infty} S(E-v)\{\exp[-\sigma(E)t] -1\}dE + n_b$$

For a thin absorber (the argument of the exponential small compared to unity), we can replace the exponential by the first terms of the series expansion and the integral in eq. (2) can be simplified to

$$(3) \quad n_e(v) = n_b - fn_0 \int_{-\infty}^{+\infty} S(E-v)\sigma(E)tdE$$

The difference between the actual spectral shape given by eq. (2) and the ideal spectrum of eq. (3) is the "blackness" effect. In the usual case that the source shape function is well approximated by a peak of Lorentz shape, and the absorption cross section can be expressed as a sum of Lorentzians, the observed spectrum for a thin absorber can be written as a sum of Lorentzians, since the convolution of two Lorentzians is also a Lorentzian. . Our transformed spectrum, now being essentially only the absorption cross section (σ) ,generally has this convenient property of being a sum of Lorentzians, even when the absorber is not thin and the source shape is not ideal.

In order to extract $\sigma(E)$ from the integral in equation (2), we must simplify and transform the expression somewhat. First we must determine n_a experimentally. This may usually be done by an additional measurement with a suitable absorber in the gamma ray path, which stops the gamma rays of the Mössbauer transition, but permits the higher energy gammas to pass through. The number of counts per channel in the background region is thus reduced from n_b to n_a, and we can calculate n_0. For the case of Co^{57} in a Cr matrix, a nickel absorber 0.005'' thick is suitable. It absorbs about 0.996 of the 14 Kev radiation, and only about 0.02 of the 122 Kev rays. We now can subtract n_b from both sides of eq. (2) and divide by fn_0, to obtain

(4) $n \equiv (n_e - n_b)/fn_0 = S(v) * \{exp -[\sigma(v)t] -1\}$

At this point our spectrum has been converted to the form of a convolution of two functions, one of which is our source shape, which we presume known. (If necessary, it can be obtained by a separate measurement with a very thin, narrow line, absorber.) Instrumental effects such as vibration broadening are included in the source shape function.

The deconvolution of an expression such as this is a well known problem in other fields; in the X-ray diffraction field it is known as the

Stokes correction (4,5). The application to Mössbauer spectra has been reported by Stone (6), but only for the purpose of sharpening the spectrum, and not for removing the blackness effect. Several numerical techniques exist for carrying out the deconvolution, but the most satisfactory seems to be the use of the Fast Fourier Transformation (FFT). (7,8,9,10)

We now carry out a finite Fourier transformation of both sides of the expression and use the property that the transform of a convolution is the product of the transforms, to obtain

$$(5) \quad \mathcal{F}\{n\} = \mathcal{F}\{S\} \cdot \mathcal{F}\{\exp[-\sigma(v)t]-1\}$$

In this transform space, we divide the transformed data by the transform of the source shape. We then antitransform to obtain

$$(6) \quad \mathcal{D}\{n\} = \exp[-\sigma(v)t] - 1$$

where $\mathcal{D}$ stands symbolically for an operator that represents the whole process of deconvolution. The procedure for carrying out this operation will be discussed in more detail later.

Finally the absorption cross section $\sigma(v)$ may be obtained by taking the logarithm of the function   .

$$(7) \quad -\sigma(v)t = \ln(\mathcal{D}\{n\}+1)$$

Figure 1 illustrates the numerical calculations used for the deconvolution of a 256 channel experimental spectra with a normalized Lorentzian of slightly more than natural width as a source shape. Figure 1 (f) shows the ratio of

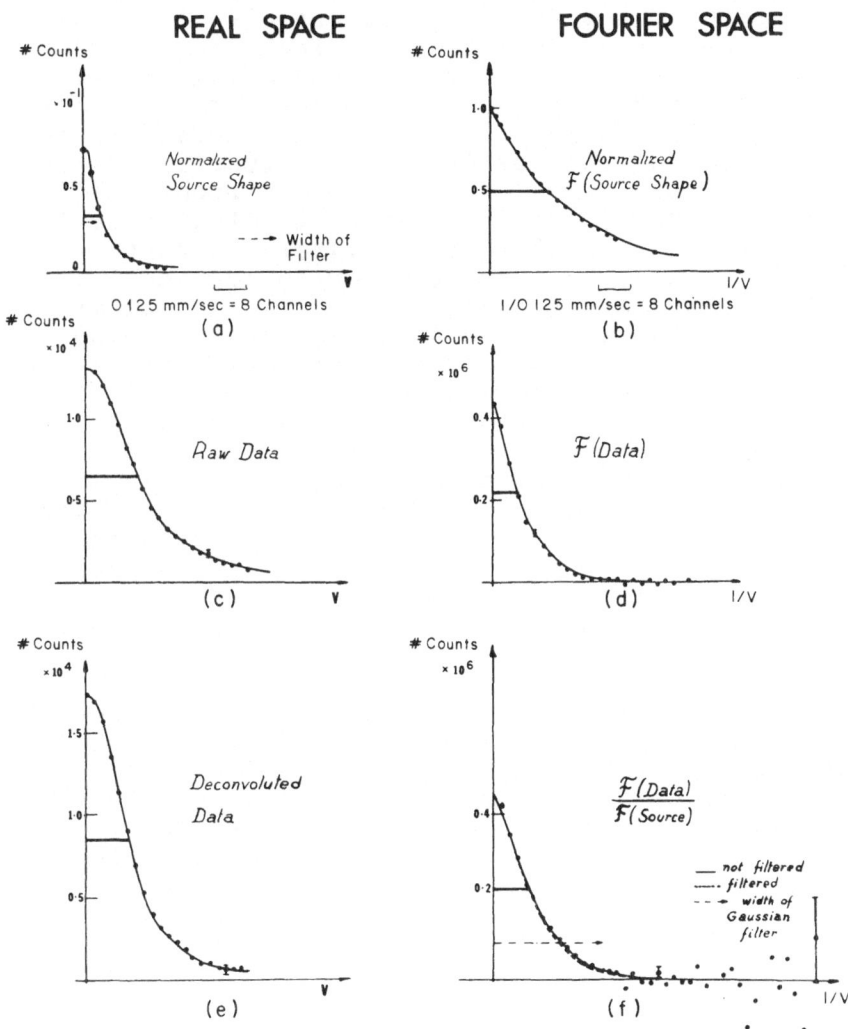

Fig. 1. Sketch to show the numerical calculation and problems involved in the deconvoluting process for a 256 channel experimental spectrum (c) with a normalized Lorentzian source shape (a). Only some points of the central region are shown. All half widths are indicated by thick lines. The filter used was a Gaussian.

the two transforms. We note that the function is smooth at low frequencies, but with increasing frequency (the independent variable in Transform space), random scatter of increasing amplitude appears, and eventually goes off scale. This scatter is a consequence of the "noise" in our data, due to the statistical fluctuations of the counting rate. It is a well known problem (11), and its effects can largely be eliminated by the use of a suitable "filter"; that is, multiplication by a function which is essentially unity at low frequency, but falls smoothly towards zero in the frequency range of high noise amplitude. The procedure for calculating an optimum filter has been given by Wiener (12), but we found a simple approximation to be adequate for our purposes. The velocity increment per channel for our spectra is small relative to the natural linewidth, so that changes in counting rate at frequencies comparable to one or two channel steps are due almost entirely to statistical "noise" and can be filtered out. We used a Gaussian filter, whose width is indicated in Figure 1 (f), whose transform in real space has a width at half maximum of about four channels, as sketched in Figure 1 (a). The use of this filter is roughly equivalent to smoothing the original data by averaging each data point with its neighbouring points. The final result of the procedure is shown as Figure 1(e). The procedure is obviously only applicable to cases where the channel spacing is small compared to the linewidth, and where the statistics are sufficiently good that adjacent channel averaging produces adequate smoothing.

The listings of the programs used are given as an appendix. The running time for a Fast Fourier Transform of 512 points is about 13 seconds in an IBM 360/67, a time comparable to the time taken by regular fitting procedures.

Figures 2 through 5 show the application of this method to several experimental spectra. In all cases the source shape used was Lorentzian

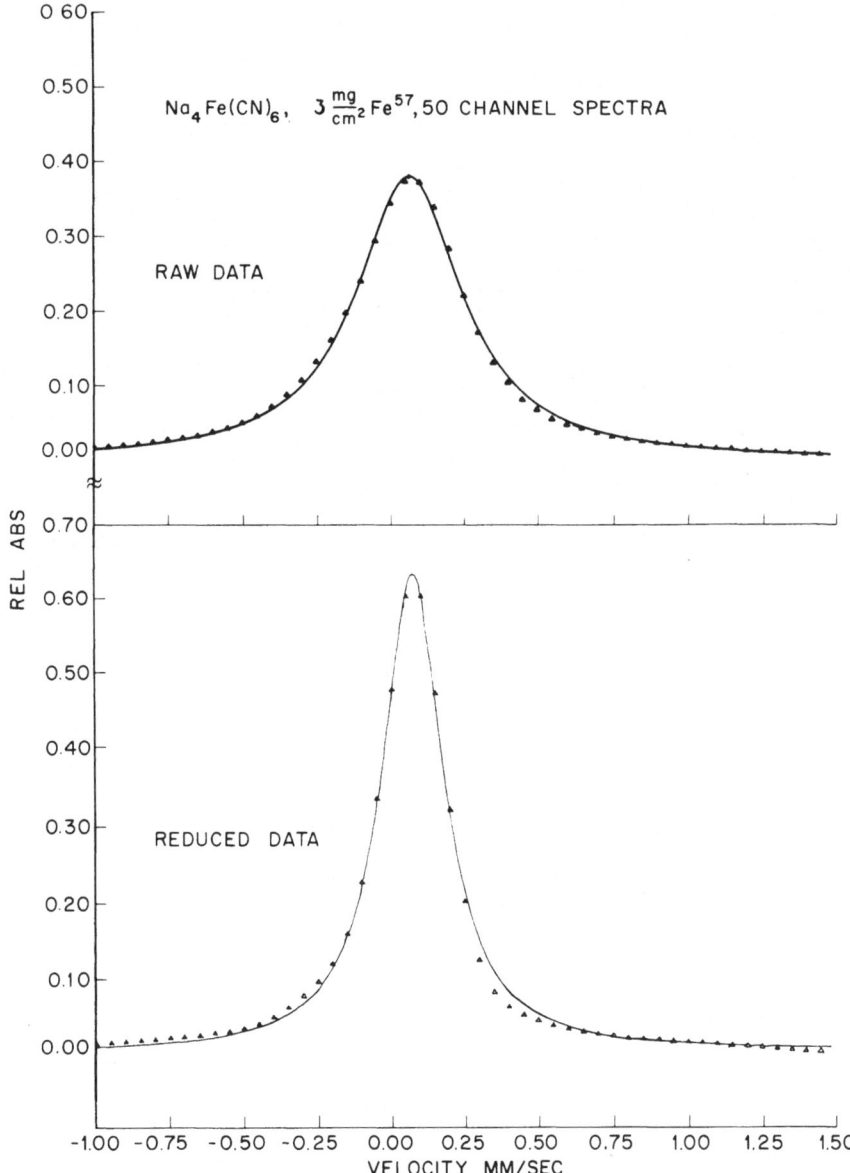

Fig. 2. Source width: 0.13 mm/sec. Velocities are taken relative to Cr source. Approximately 5.10^6 counts per channel. The relative absorption in the reduced data should be multiplied by a factor of 2 to obtain σ.

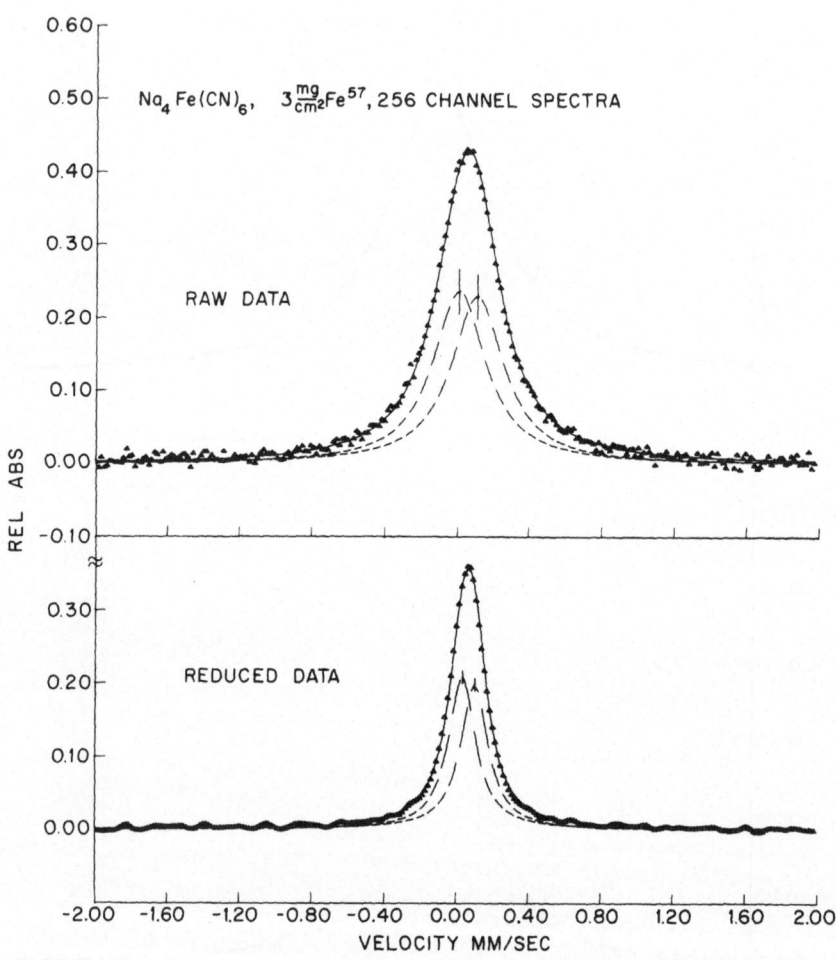

Fig. 3. Source width: 0.13 mm/sec. Velocities are taken relative to Cr source. Approximately 3.10^4 counts per channel. The relative absorption in the reduced data should be multiplied by a factor of 5 to obtain σ.

with a width of 0.0136 mm/sec, slightly greater
than natural. This value was based on
measurements with a very thin, high purity iron
foil, and the assumption that the inner peaks
were of natural width.

Figure 2 shows a 50 channel spectrum of
$Na_4Fe(CN)_6$ before and after processing. The raw
spectrum has a very smooth background and the
processing does not result in any noticeable
change. This material is generally regarded as a
single line absorber, and, in this figure, one
peak fits are shown. It is important to note
that the processing we describe is quite
independent of the fitting procedure. It is
often of interest to carry out a conventional
least squares fitting to a sum of Lorentzians for
that data, both before and after transformation,
but that is a completely independent calculation.
Figure 3 shows spectra for the same material,
taken with a 256 channel analyser, and in this
case, fitted to a sum of two Lorentzians. A
slightly better fit is obtained with two peaks
than with one, a result consistent with the work
of Evans and Black, who established with single
crystal measurements that quadrupole splitting is
definitely present in this material. (13) In
this and the following figures, the effect of the
filter on smoothing the background is evident.
Note that each high point in the background of
the transformed spectrum corresponds to a group
of high points in the raw data.

Figures 4 and 5 show spectra from a material
with a clearly resolved quadrupole splitting,
$K_3FE(CN)_6$. The sample thickness was changed by a
factor of 22 between the two samples, and the
blurring effect of thickness is clearly seen in
the raw data. In the case of the reduced data,
however, the spectra have the same shape; the dip
between the peaks is about half the peak
height.

The effect of this blackness correction
procedure on peak widths is shown in Figure 6,
where the peak widths obtained for the two

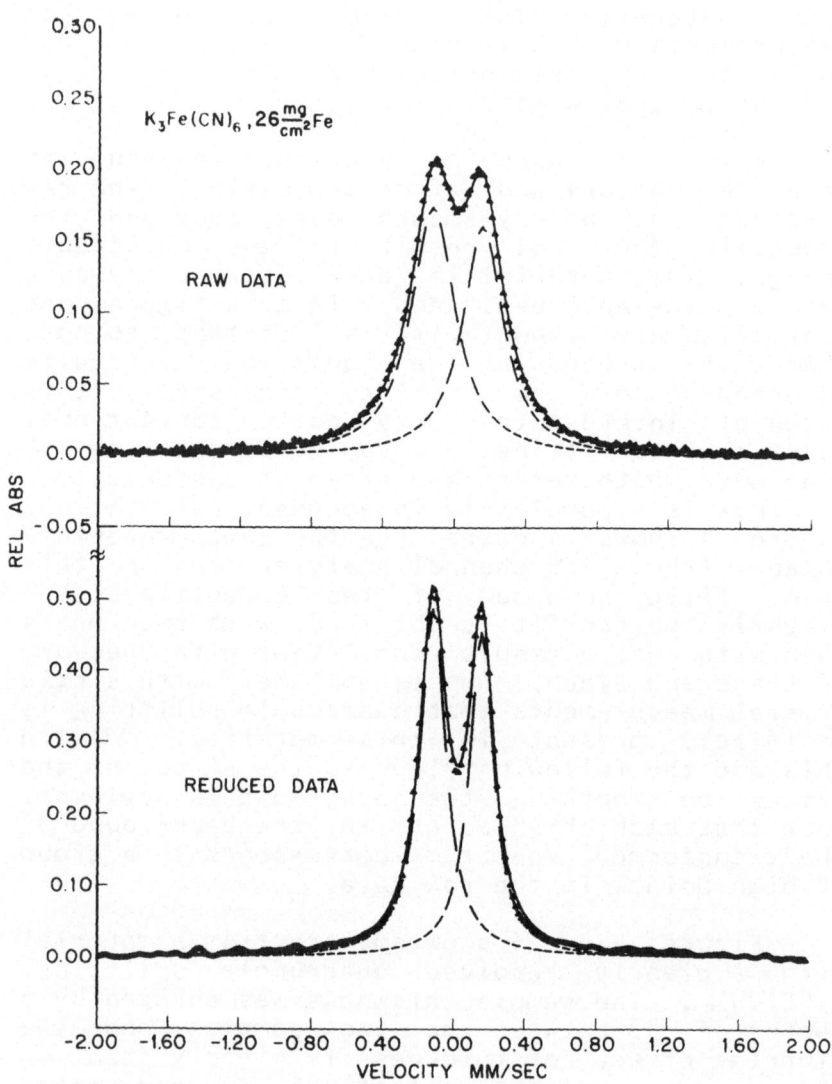

Fig. 4. Source width: 0.13 mm/sec. Velocities are taken relative to Cr source. Approximately 5.10^5 counts per channel. The relative absorption in the reduced data should be multiplied by a factor of 5 to obtain σ.

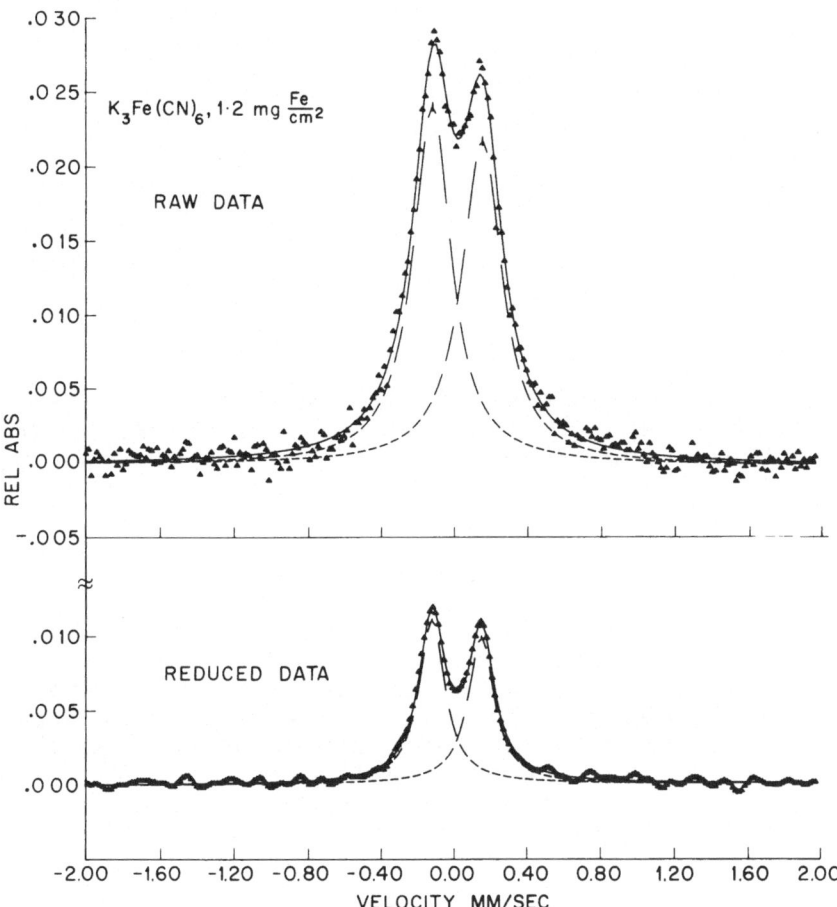

Fig. 5. Source width: 0.13 mm/sec. Velocities
are taken relative to Cr source. Approximately
3.10^5 counts per channel. The relative absorption
in the reduced data should be multiplied by a
factor of 5 to obtain σ.

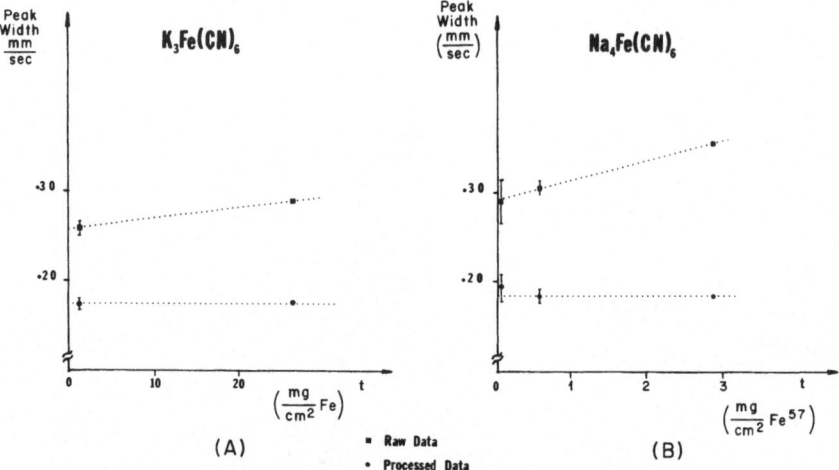

Fig. 6. Peak widths (full widths at half maximum) before and after processing, for $Na_4Fe(CN)_6$ and $K_3Fe(CN)_6$. Fit to quadrupole doublets in both cases. Source used : Lorentzian of width=0.13mm/sec.

samples by least square fitting of both raw and transformed data are plotted. The usual dependence of peak width on sample thickness (2,3) is observed for the peak widths of raw data. In both cases, the width obtained from the transformed data is about 0.17 mm/sec, made up of a natural width of 0.1 mm/sec and a contribution from the filter of about 0.07 mm/sec. More important than the exact value obtained for the peak width is the fact that is independent of sample blackness, so that additional contributions to the broadening due to diffusion, for example, can be determined unambiguously.

In addition to removing the effects of sample blackness on peak width, this procedure also removes the distortion of peak amplitudes and areas. This is important in such cases as the use of Mössbauer spectroscopy for the quantitative analysis of multiphase mixtures. After transformation, the areas of the portions of the spectrum due to each constituent are simply proportional to the amount of each phase present multiplied by the appropriate Debye-Waller factor. A similar problem arises in the case of the spectrum of a solid solution, where various local enviroments give rise to different components of the overall spectrum. A quantitative determination of the relative numbers of Mössbauer active atoms in each kind of environment again requires removal of the thickness effect. Another related problem is the investigation of complicated magnetic structures. Unless the sample is ideally thin, or the blackness effect is removed, the areas under the peaks are not proportional to the true intensities.

ACKNOWLEDGMENTS

The authors are grateful for helpful discussions with Dr. W. Flanagan, Dr. R. Marburger and Dr. F. de S. Barros.

APPENDIX

```
C PROGRAM - BLACKCOR

C MAIN PROGRAM TO REMOVE SOURCE EFFECT AND
C BLACKNESS DISTORTION
        COMPLEX B,C,X2
        DIMENSION B(2000),C(2000),X(2000),X2(2000)
        DIMENSION FMT(18),LABEL(18)
C VARIABLES USED:-
C INPUT:  ARRAY X OF EXPERIMENTAL DATA WITH
C VARIABLE FORMAT
C         ARRAY C OF TRANSFORMED, FILTERED
C SOURCE  SHAPE (EG. LORGEN PROGRAM FOR
C GENERATION OF LORENTZIAN SOURCES)
C          B ARRAY REPRESENTS COMPLEX DATA,
C COMPLETED WITH ZEROS TO NL IN
C NUMBER (IT IS NECESSARY TO HAVE AT LEAST
C AS MANY ZEROS AS DATA POINTS TO CARRY
C OUT THE CONVOLUTION CORRECTLY, SO WE
```

```
C TAKE IT TO BE THE SMALLEST POWER OF
C TWO BIGGER THAN TWICE THE NUMBER OF
C DATA POINTS,WHICH ENABLES US TO USE THE
C SIMPLEST FORM OF FAST FOURIER TRANSFORM
C ALGORITHM).
C
C X2 IS THE DATA IN FOURIER SPACE
C
C INPUT AND PREPROCESSING OF DATA
C
      WRITE(5,5)
 5    FORMAT('  ENTER  NUMBER  OF  EXPERIMENTAL
POINTS AS I4')
      READ(5,10)N
 10   FORMAT(I4)
      NSTAGE=3
 30   NSTAGE=NSTAGE+1
 75   IF(2*N .GT. 2**NSTAGE)GO TO 30
 20   WRITE(5,33)
 33   FORMAT(' RESONANT FRACTION?')
      READ(5,40)F
 40   FORMAT(F10.8)
      WRITE(5,50)
C THE BACKGROUND SHAPE USED IN THIS PROGRAM IS
C BKG(1+A*DISTANCE TO CENTER OF SPECTRUM**2)
 50   FORMAT('  ENTER  BACKGROUND  PARAMETERS:
CONSTANT  BKG,NUMBER  OF  122S  AND  NORMALIZED
CURVATURE AS 2F12.4,F12.10')
      READ(5,60)BKG,RN122,A
 60   FORMAT(2F12.4,F12.10)
      NL=2**NSTAGE
      WRITE(5,70)
 70   FORMAT('  ENTER  FORMAT  TO  BE  USED  IN
READING DATA')
      READ(5,12)FMT
 12   FORMAT(18A4)
      READ(8,FMT)(X(I),I=1,N)
      FND2=FLOAT(N)/2.
      FBR=F*(BKG-RN122)
      DO 190 L=1,N
      BCOR=1.+A*(FLOAT(L)-FND2)**2
 190  B(L)=CMPLX((X(L)/BCOR-BKG)/FBR,0.0)
      N1=N+1
      DO 200 L=N1,NL
 200  B(L)= CMPLX(0.0,0.0)
C
```

```
      CALL FASTF(NSTAGE,+1.,B,X2)
C
C DIVISION IN FOURIER SPACE
C
      READ(7,500)( C(I),I=1,NL)
  500 FORMAT(2(G16.7,4X))
      DO 600 I=1,NL
  600 B(I)=X2(I)/ C(I)
C
      CALL FASTF(NSTAGE,-1.,B,X2)
C
C THICKNESS CORRECTION
C
      WRITE(5,700)
  700 FORMAT(' ENTER LABEL')
      READ(5,12)LABEL
C
C BL= 1.0 IS ADDED SO THE FITTING PROGRAM WILL
C COMPARE NUMBERS WITH UNITY
      BL=1.
      DO 800 I=1,N
  800 X(I)=ALOG(REAL(X2(I))+1.0)+BL
      WRITE(10,12)LABEL
      WRITE(10,350)(X(I),I=1,N)
C OUTPUT HAS BEEN REDUCED TO ORIGINAL NUMBER OF
C POINTS,AND ONLY THE REAL PART IS TAKEN
C (THE IMAGINARY PART SHOULD BE ZERO EXCEPT
C FOR ROUNDOFF ERRORS )
  350 FORMAT(5G16.7)
      STOP
      END



C PROGRAM - LORGEN

C PROGRAM TO GENERATE A COMPLEX,CENTERED
C LORENTZIAN,TRANSFORM IT BY MEANS OF THE FAST
C FOURIER TRANSFORM AND FILTER IT WITH A
C GAUSSIAN.
C THE NUMBER OF POINTS SHOULD BE AT LEAST TWICE
C THE NUMBER OF EXPERIMENTAL DATA POINTS WHICH
C ARE BEING PROCESSED BY THE MAIN PROGRAM
C AND MUST BE A POWER OF TWO.
C SUBROUTINE REQUIRED=FASTF
C INPUT=NUMBER OF STAGES(N=2**NSTAGE),
```

```
C HALF WIDTH OF THE LORENTZIAN AND FILTER
C CONSTANT,FROM WHICH THE WIDTH OF THE
C GAUSSIAN FILTER IS DETERMINED AS THE ORDINATE
C OF THE LORENTZIAN TRANSFORM WITH THAT VALUE.
C INITIAL VARIABLES:G IS THE ARRAY FOR THE
C GENERATED LORENTZIAN,F FOR THE CENTERED,
C COMPLEX LORENTZIAN,X FOR THE TRANSFORMED
C LORENTZIAN,DIV TO SEARCH FOR FILTER WIDTH.
        COMPLEX F,X
        DIMENSION G(2000),F(2000)
        DIMENSION DIF(2000),X(2000)
        WRITE(5,10)
  10    FORMAT(' ENTER   NSTAGE(N=2**NSTAGE)     AND
HALFWIDTH AS I4,F8.3')
        READ(5,20)NSTAGE,A
        N=2**NSTAGE
  20    FORMAT(I4,F8.3)
        ND2=N/2
        NMIN=-ND2-1
        ADPI=A/3.141593
        DO 100 I=1,N
        XG=FLOAT(NMIN+I)
 100    G(I)=ADPI/(A*A+XG*XG)
        DO 200 I=1,ND2
        K=I+ND2
 200    F(I)=CMPLX(G(K),0.0)
        DO 300 K=1,ND2
        I=K+ND2
 300    F(I)=CMPLX(G(K),0.0)
        CALL FASTF(NSTAGE,+1.,F,X)
        WRITE(5,500)
 500    FORMAT(' ENTER FILTER CONSTANT')
        READ(5,600)D
 600    FORMAT(F10.8)
        ND21=ND2+1
        XM=10.
        DO 700 I=1,ND21
        DIF(I)=CABS(X(I)-D)
        IF(DIF(I) .GT. XM)GO TO 700
        XM=DIF(I)
        J=I
C J IS THE FILTER WIDTH IN CHANNELS
 700    CONTINUE
        SJ=FLOAT(J)*FLOAT(J)
        DO 800 I=1,ND21
        FI=FLOAT(I-1)
```

```
C ARRAY F IS NOW THE FILTER
      F(I)=EXP(-0.50*FI*FI/SJ)
 800  X(I)=X(I)/F(I)
      DO 900 K=2,ND2
      I=N-K+2
      F(I)=F(K)
C X IS NOW FILTERED,TRANSFORMED LORENTZIAN
 900  X(I)=X(I)/F(I)
C THE OUTPUT X IS COMPLEX
      WRITE(8,950)(X(I),I=1,N)
 950  FORMAT(2(G16.7,4X))
      STOP
      END



C SUBROUTINE FASTF

C PERFORMS FAST FOURIER TRANSFORMS
C CALLING PARAMETERS:NSTAGE (NUMBER OF POINTS
C OF INPUT ARRAY=2**NSTAGE),SIGN GIVES THE
C DIRECTION OF THE TRANSFORM, X1 AND X2 ARE
C THE COMPLEX INPUT AND OUTPUT ARRAYS
      SUBROUTINE FASTF(NSTAGE,SIGN,X,X2)
      COMPLEX X,X1,X2,W
      DIMENSION X(2000),X1(2000),X2(2000)
      PI2=6.2831853
      NL=2**NSTAGE
      DO 25 I=1,NL
  25  X1(I)=X(I)
      INTEGER R
      N2=NL/2
      FLTN=FLOAT(NL)
      DO 3 K=1,NSTAGE
      N2K=NL/(2**K)
      NR=N2K
      N1=(2**K)/2
      DO 2 I=1,N1
      IN2K=(I-1)*N2K
      FLIN2K=IN2K
      SAN=FLIN2K/FLTN
      ASAN=AMOD(SAN,1.)
      TEMP=PI2*SIGN*ASAN
      W=CMPLX(COS(TEMP),SIN(TEMP))
      DO 2 R=1,NR
      ISUB=R+IN2K
```

```
      ISUB1=R+IN2K*2
      ISUB2=ISUB1+N2K
      ISUB3=ISUB+N2
      X2(ISUB)=X1(ISUB1) +W*X1(ISUB2)
      X2(ISUB3)=X1(ISUB1) -W*X1(ISUB2)
2     CONTINUE
      DO 3 R=1,NL
3     X1(R)=X2(R)
      IF(SIGN.GT.0.)GO TO 5
      DO 4 R=1,NL
4     X2(R)=X1(R)/FLTN
5     CONTINUE
      RETURN
      END
```

REFERENCES

1.   J. Heberle and S. Franco, Z. Naturforsch, 10, 1439 (1968).

2.   S. Margulies and J. R. Ehrman, Nuclear Instr. and Meth. 12, 131 (1961).

3.   S. L. Ruby and J. M. Hicks,Rev. Sci. Instr. 33, 27 (1962).

4.   J. W. M. Du Mond, Revs. Modern Phys., 5, 1 (1933).

5.   W. A. Rachinger, J. Sci. Instr., 25, 254 (1948).

6.   A. J. Stone, Chem. Phys. Letters 6,331 (1970).

7.   J. W. Cooley, P. A. Lewis and P. D. Welch, IEEE Trans. Audio Electroacoust. AU17, 77 (1969).

8.   W. T. Cochran et al., Proc. IEEE 55, 1664 (1967).

9.   G. D. Bergland, IEEE Spectrum, 42 (July 1969).

10. M. L. Uhrich, IEEE Trans. Audio Electroacoust. AU17, 170 (1969).

11. J. S. Rollet and L. A. Higgs, Proc. Phys. Soc., 79, 87 (1962).

12. N. Wiener,"Extrapolation, Interpolation and smoothing of Stationary Time Series", Technology Press of MIT and John Wiley & Sons, Inc., New York, (1950).

13. M. J. Evans and P. J. Black,Proc. Phys. Soc. 3, L81 (10).

THE EFFECTS OF RADIO FREQUENCY FIELDS ON FERROMAGNETIC
MÖSSBAUER ABSORBERS

Loren Pfeiffer

Bell Telephone Laboratories, Incorporated

Murray Hill, New Jersey   07974

A class of experiments is discussed in which the
Mössbauer Effect is used as a probe to observe the effect
of radio frequency fields on iron containing materials.
These experiments have so far led to the discovery of two
new physical effects.

One of the effects was discovered by accident when it
was noticed that extra absorption lines were appearing in
the spectrum of an iron absorber foil subjected to an rf
field of a few Gauss. It quickly became apparent that
these additional lines were in fact satellites occurring
symmetrically on both sides of each of the six unperturbed
iron lines, and that the satellites were displaced from
the main lines by multiples of the applied frequency. A
review of our current understanding of this rf sideband
effect is given together with some recent data showing the
effect at very high applied frequencies, and as a function
of several magnetic phase transitions.

The second rf effect was predicted on the basis of a
simple model but was not actually seen until after some
considerable experimental effort was expended. The model
involved extending the technique of rf hyperfine enhancement to the extreme, so that the entire magnetic hyperfine
field at an iron nucleus would be forced to oscillate in
time through large angles in response to an applied rf
driving field. If this were done, and if the driving
frequency were made several times the nuclear Larmor
frequency of the iron, then the time average field as seen

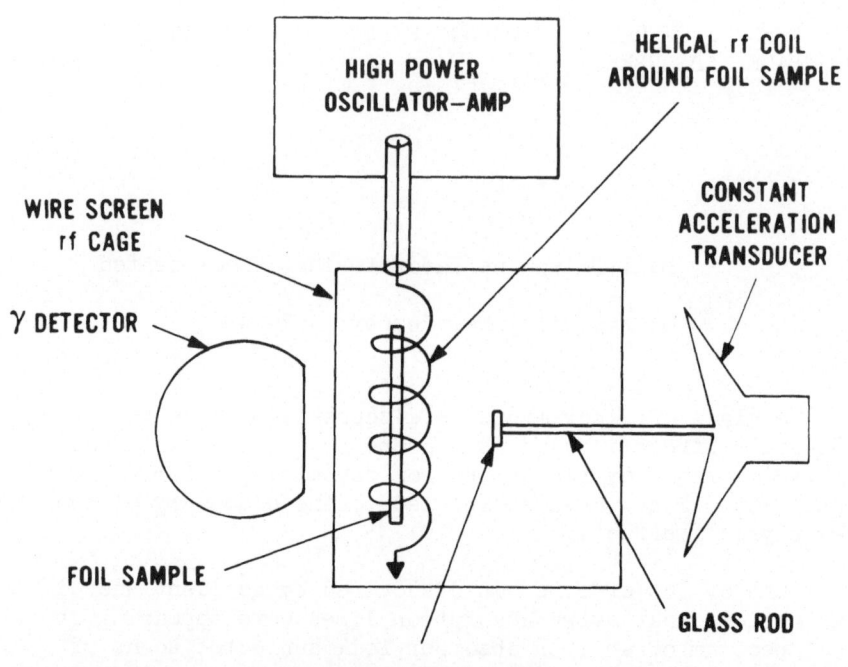

Co^{57} SOURCE IN Pd MATRIX

Fig. 1   Schematic Diagram of Experimental Apparatus. This
equipment was used to observe both the rf sideband and the
rf collapse effects. The constant acceleration transducer,
Co^{57} in Pd source, absorber foil and gamma detector together
form a conventional Mössbauer effect spectrometer which is
used to study the Fe^{57} nuclei in the absorber foil. A
linearly polarized rf magnetic field is applied to the absorber by means of the helical wire coil and the high power
oscillator amplifier. The rf oscillator is a General Radio
1164A frequency synthesizer and the rf amplifier is an
Instruments for Industry model 404A. The rf field is measured
by monitoring the current through the helical coil with an
rf thermocouple ammeter. The screen cage surrounding the
absorber helix is necessary to prevent radiated rf power
from interfering with the other electronic equipment.

by the iron nuclei would be reduced from its original value
of ∿300KGauss to zero, and the entire six-line Mössbauer
absorption spectrum would collapse together to become a
single central line. This rf collapse effect was recently
observed in a foil of permalloy. This data together with
other supporting evidence for the effect is discussed.

INTRODUCTION

For the past few years, I have been using the Mössbauer
effect as a probe to observe the effects of radio frequency
magnetic fields on iron nuclei in various materials. The
work has been exploratory in spirit, and thus far has led
to the discovery of two new effects.

The first effect, called "rf Mössbauer sidebands" was
discovered in 1968 while I was at the Johns Hopkins
University. The original work was done there (1) in col-
laboration with J. C. Walker and Neil Heiman. In the first
part of this paper we will discuss some recent experiments
which have helped to better understand the mechanism of the
sideband effect.

The second part of the paper will be a discussion of a
rather startling effect that was first observed only a few
months ago at Bell Laboratories (2). I have called this
new effect rf collapse of the hyperfine field.

rf SIDEBAND EXPERIMENTS

The apparatus that one can use for observing either
effect is shown in the first figure. The equipment is quite
straightforward. It is built around a conventional constant
acceleration Mössbauer spectrometer in transmission geometry.
The recoilless 14.4 keV gamma rays from the Co^{57} in Pd source
pass through the fixed absorber under study, and are detected
using a Kr filled proportional counter and conventional
electronics.

The differences from a conventional set up are that
there is a helical coil of silver wire wrapped around the
Mössbauer absorber so that a radio frequency magnetic field
can be applied in the plane of the absorber, and that there
is a screen cage to shield the electronics from possible rf
interference.

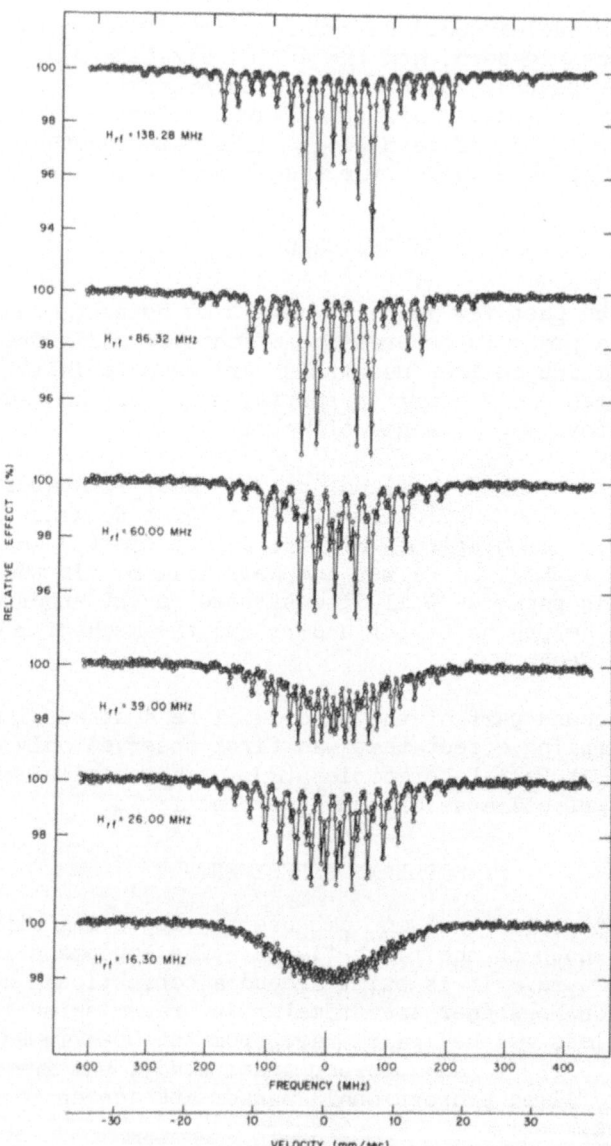

Fig. 2   The rf Mössbauer Sideband Effect as a function of
driving frequency. For each Mössbauer spectrum the rf magnetic
field was 7.5 Oe in peak amplitude directed in the plane of
the absorber. The absorber was a foil of 99.99% purity natural iron 25 μm thick and 12.5 mm in diameter, which had been
previously annealed in dry hydrogen at 950°C for several hours.

Now if one puts a natural metallic iron foil 25 μm thick into the apparatus as the absorber, and adjusts the rf circuitry so that a 7.5 Oe magnetic field oscillating at 138.28 MHz is applied in the plane of the foil, then one obtains the Mössbauer spectrum shown in the top trace of Fig. 2.

Clearly something not altogether expected has happened. The usual six absorption lines of metallic iron are found at their proper velocities in the center of the pattern, but there are additional absorption lines as well. These additional lines are in fact satellite sidebands occurring in a series of symmetric pairs of diminishing intensity on either side of each of the six unperturbed iron lines.

To determine the location of the sidebands quantitatively one must know the iron splittings in absolute frequency units. The Fe^{57} ground state splitting is known from NMR measurements to be 45.49 MHz at 300°K (3). From this and the knowledge of the excited state g factor obtained from Mössbauer measurements, one can calculate that the excited state splitting is 26.0 MHz, and that the splitting of the outer lines is 123.5 MHz.

If one now fits the sideband spectra of Fig. 2 using these calibration frequencies one finds that the sidebands are displaced from their parent lines by exact multiples of the applied frequency, that is by $\pm n\omega_m$. The smooth curves in Fig. 2 show the excellent fits for various applied frequencies that one obtains with this simple theory.

For ease of comparison I have labeled the energy axis of Fig. 2 in both the common velocity units used by most workers in Mössbauer spectroscopy, and also in absolute frequency units. The frequency units are actually much more convenient in this work as one can displace the sidebands from their parents by any amount in frequency by just dialing in that frequency on the rf generator.

The rf sideband effect is in fact very useful as a Mössbauer calibration technique. Calibration using rf sidebands has the great advantage of giving the spectroscopic energy splittings directly in frequency units, thereby circumventing all the difficulties involved in accurately measuring the source velocity and the subsequent conversion to energy units. The sideband method is also useful in

determining the velocity linearity of a Mössbauer spectrometer, since as will be shown later it can be used to
generate a calibration picket fence of absorption lines each
of which is separated from its adjacent neighbors by the
same precisely known frequency.

Granted that the effect is useful; what is the mechanism
that causes it to occur? First let me be clear about what is
not the mechanism of the effect. The sidebands we are discussing are not the sort of thing seen by Ruby and Bolef (4)
in 1960. These workers and others more recently (5,6) mount
their samples on quartz transducers which are set into
piezoelectric vibration by an rf electric field. There is
no quartz transducer in this experiment. This is just a rf
field on a magnetic foil giving sidebands, so that we need
some other mechanism.

The mechanism I wish to propose is that there is a
magneto-acoustic coupling in the iron foil itself, that this
coupling causes a Doppler motion of the iron-absorbing
nuclei, and that this motion results in an apparent frequency
modulation of the gamma ray which produces the sideband absorption lines. I will now give a very simple classical
picture of how the Doppler motion produces the sidebands.
A more rigorous quantum treatment is being published elsewhere (7), but I believe this treatment is of value because
it gives one an intuitive feeling for the physics.

FREQUENCY MODULATION OF GAMMA RAYS

We may write the time dependent electric vector, $E(t)$
of an unmodulated gamma ray of frequency ω_0 as

$$\vec{E}(t) = \vec{E}_o \, e^{i\omega_0 t} .$$

Now assume that some mechanism as yet unspecified, causes
each nucleus in the Mössbauer absorber to vibrate with peak
amplitude x_0 and frequency ω_m, then

$$x(t) = x_o \sin \omega_m(t) .$$

Now from the point of view of the moving nuclei, the gamma
ray incoming along the x axis will be Doppler shifted in
frequency to

$$\omega(t) = \omega_o\left(1 + \frac{1}{c}\frac{dx}{dt}\right)$$

$$= \omega_o\left(1 + \frac{x_o\omega_m}{c}\cos\omega_m t\right).$$

Thus the electric vector of the gamma ray becomes

$$\vec{E} = \vec{E}_o \exp i \int\omega(t)dt$$

$$= \vec{E}_o\exp\left(i\omega_o t + \frac{i\omega_o x_o}{c}\sin\omega_m t\right).$$

This last expression may be expanded as a Fourier Series by using the Bessel Function identity (8),

$$e^{ia \sin b} \equiv \sum_{n=-\infty}^{\infty} J_n(a)e^{inb},$$

where $J_n(a)$ is the nth unmodified Bessel function of the first kind with argument a. Performing this substitution and rearranging factors results in an expression for the gamma ray electric vector which explicitly shows the sideband components.

$$\vec{E}(t) = \sum_{n=-\infty}^{\infty} J_n\left(\frac{\omega_o x_o}{c}\right)\vec{E}_o\, e^{i(\omega_o+n\omega_m)t}.$$

Now the intensity of a Mössbauer absorption line goes as the square of the gamma ray electric vector, so that the intensity of the nth sideband will scale as the square of the nth Bessel coefficient,

$$J_n^2\left(\frac{x_o}{\lambda}\right). \tag{1}$$

The argument of the Bessel coefficient above is called the modulation index. In writing the argument I have simplified the notation replacing c/ω_o by λ, the normalized wavelength of the gamma ray. This wavelength λ for 14.4 keV Fe^{57} gamma

rays is numerically equal to 0.137 Å. Thus from the properties of the J_n we see that numbers of rf sidebands of large
intensity will occur in the Mössbauer spectrum, if the
absorber nuclei are set into vibration at a peak amplitude
of the order of 0.137 Å.

In the discussion just given the simplifying assumption
was made that all of the absorber nuclei vibrate with the
same peak amplitude x_0. If, however, the peak amplitude x_0
is not constant, but has a distribution $w(x_0)$ over the sample
volume, then the sideband intensities will scale as

$$\int_0^\infty w(x_0)\ J_n^2\left(\frac{x_0}{\lambda}\right) dx_0 \ . \tag{2}$$

This quite general expression for the rf sideband intensities
is identical with the result derived quantum mechanically (7).

The unsolved problem of course is to specify the form
of $w(x_0)$, the peak amplitude distribution function. Allowing $w(x_0)$ to be a Dirac delta function, for example,
reduces expression (2) to the form of expression (1).
Expression (1) is not, however, in agreement with the experimentally observed sideband intensities. At a modulation
index of 2.41, for example, the unshifted parent line corresponding to J_0^2 is predicted to have zero intensity; this
however is never seen experimentally.

Abragam (9) in connection with a similar problem
proposed a different distribution function. He assumed that
the nuclear Doppler motion could be written as

$$x(t) = \alpha \sin \omega_m t + \beta \cos \omega_m t$$

where the phase amplitudes α and β each have a Gaussian
distribution. In this case the resulting peak amplitude
$x_0 = (\alpha^2+\beta^2)^{1/2}$ is describable as a Rayleigh distribution,

$$w(x_0) = \frac{x_0}{a^2} e^{-\frac{x_0^2}{2a^2}} \ . \tag{3}$$

Substituting this distribution function into expression (2) and evaluating the integral gives

$$e^{-\frac{a^2}{\lambda^2}} \; I_n\left(\frac{a^2}{\lambda^2}\right) \tag{4}$$

for the relative sideband intensities, where the $I_n(m^2)$ are the modified Bessel functions of the first kind.

Unfortunately the observed experimental sideband intensities are in disagreement with this form as well. The disparity between experiment and this expression is not as violent, however, as between experiment and expression (1). In most cases the sideband intensities predicted by expression (4) are within about 20% of the measured intensities. For this reason it has been possible to use expression (4) to at least characterize the experimental sideband data with an approximate modulation index, (c.f. Fig. 10).

In Fig. 3 I have plotted both the Dirac delta function and the Rayleigh distribution function that we have considered so far. In a sense these two functions represent the extremes of a whole class of possible functions. The correct explicit form for $w(x_o)$ will hopefully be suggested by the proper model for the magneto-acoustic coupling.

EXPERIMENTAL DATA ON rf SIDEBANDS

What is in fact the mechanism which causes the nuclei in the absorber to vibrate at the frequency of the applied rf electromagnetic field? To narrow down the possibilities we must know some of the properties of the effect. On what experimental parameters do rf sidebands depend? What follows then will be a review of the experimental properties of rf sidebands.

First, sidebands occur only if the Mössbauer absorber is ferromagnetic. The proof of this is summarized in the next two figures which show the dramatic changes in sideband intensity that are seen when the sample undergoes a ferromagnetic phase transition. In Fig. 4 we see the behavior of sidebands in $FeBO_3$ single crystals subject to an applied rf

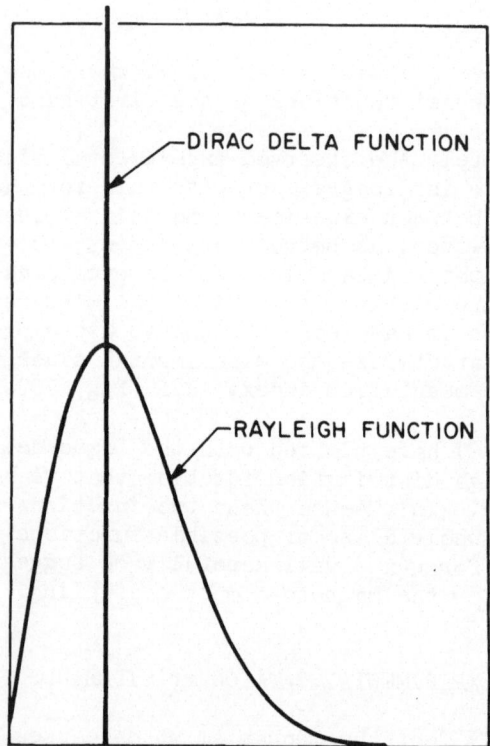

Fig. 3   Graphical representation of the Dirac Delta Function
and the Rayleigh Distribution Function which are discussed
in the text.

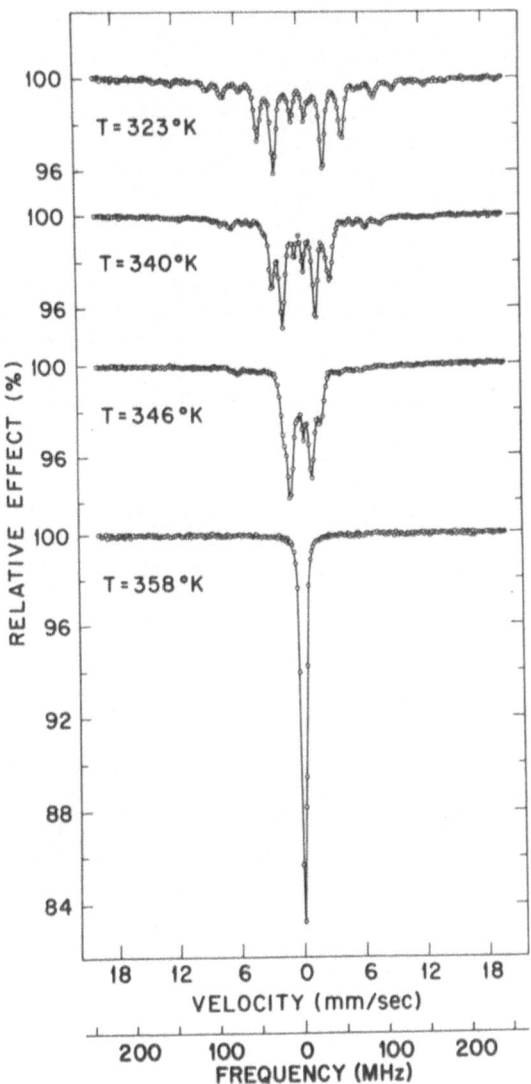

Fig. 4   The behavior of the rf sideband effect near the
Curie temperature. The rf field for all the spectra was
5.0 Oe peak amplitude at 61.0 MHz. The absorber was made
from single crystals of $FeBO_3$ which (Ref. 10) has a Curie
temperature of 348°K. The sample temperature was controlled
using a thermocouple and a conventional resistance heater
in a feedback loop as in Ref. 10.

magnetic field of 5.0 Oe at 61.0 MHz.   In these data the
applied rf field was held constant, and the sample temperature
was varied using a conventional resistance heater in a temperature feedback loop. $FeBO_3$ is a weak ferromagnet with a
Curie temperature of $348°K$, (10) and indeed one sees in the
figure that the sidebands disappear as the sample temperature
is raised above the Curie point.

   In Fig. 5 similar behavior is shown for the ferromagnetic
to antiferromagnetic Morin transition in single crystals of
Fe_2O_3.   For this case as well we see sidebands only if the
sample has a ferromagnetic moment.   Above the Morin temperature the Fe_2O_3 crystals are ferromagnets and show the sideband effect.   On the other hand below the -13°C transition
the crystals are antiferromagnets with no canting and the
sideband effect does not occur.

   Another property of the effect is that it does not
occur in powdered materials.   Sidebands are readily seen
for example in iron foils or in Fe_2O_3 single crystals, but
if these materials are broken into micron-sized particles no
rf effect is seen (1).   Possibly the reason for this behavior
is that in the small particles the applied rf field is
effectively cancelled out by demagnetizing fields which become
larger as particle size decreases.   A second possibility is
that one might expect the amplitude of the acoustic vibrations to be small in particles which are themselves small
compared to the acoustic vibration wavelength.   These possible explanations can be sorted out experimentally, and such
experiments are in fact under way.   Nevertheless, the conclusion stands that to see rf sidebands one needs a bulk
ferromagnetic sample.

   A third property of the effect is that the number of
ferromagnetic domains or the motion of domain walls are
apparently not important for sidebands.   To prove this
experimentally one sweeps away the domain walls with a large
static field $\vec{H}_0$, and monitors the behavior of the rf sidebands.

   To determine how large a sweep field $\vec{H}_0$ is required,
the magnetic hysteresis loop for our 8μ thick annealed iron
sample was run and is shown in Fig. 6.   One sees there that
fields as small as 100 Oe are large enough to insure that
the domain walls have been swept away.

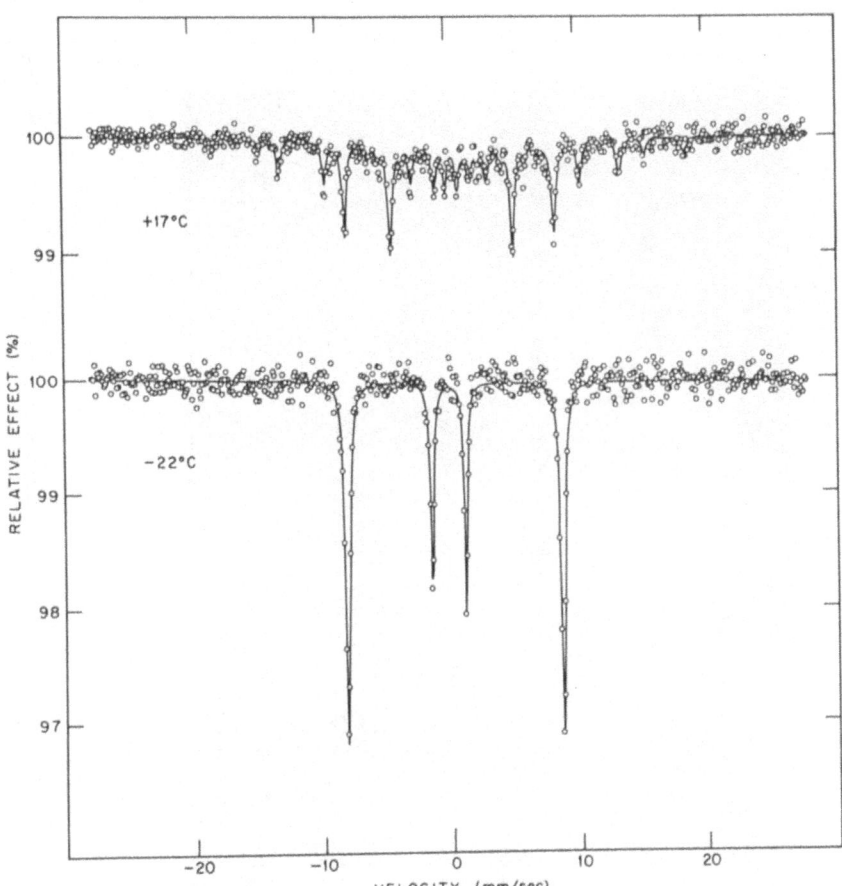

Fig. 5   The behavior of the rf sideband effect near the
Morin ferromagnetic-to-antiferromagnetic transition in
single crystals of α Fe_2O_3. The rf field for both spectra
was 22 Oe peak amplitude at 61.4 MHz. The rf sideband
effect is seen at +17°C where the Fe_2O_3 crystals are ferromagnetic, but the effect is not seen at -22°C where the
crystals are antiferromagnetic.

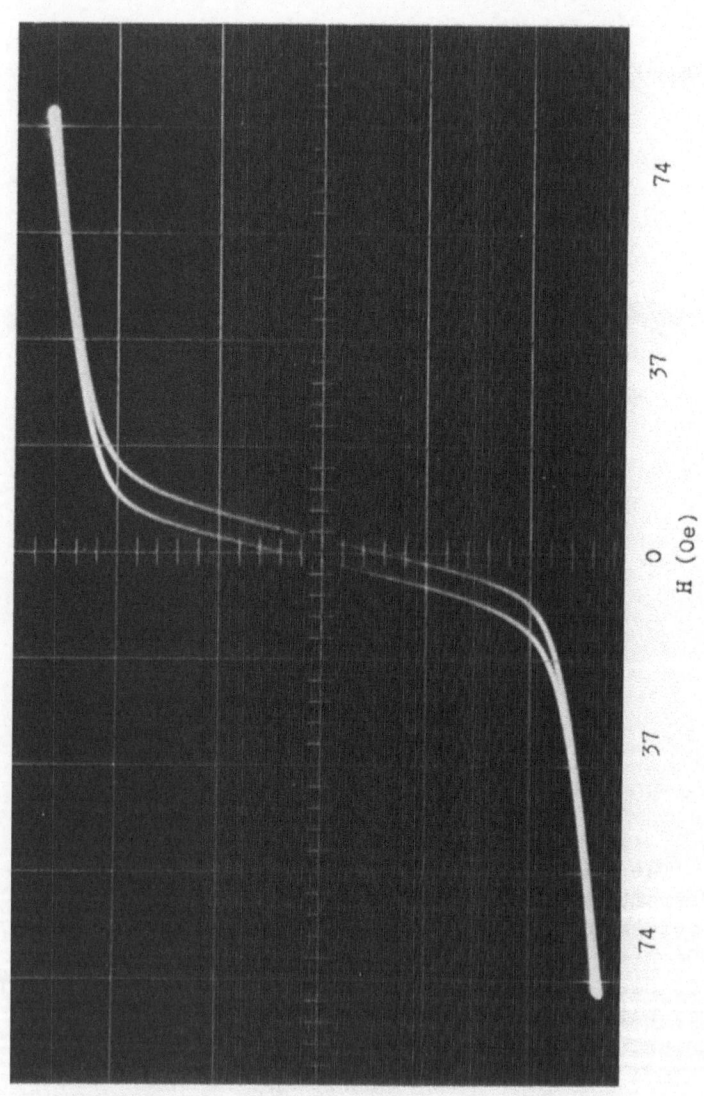

INDUCTION B

H (Oe)

Fig. 6   The $\vec{B}$ - $\vec{H}$ hysteresis loop of the 8μ thick annealed iron foil used for the static field experiments of Figs. 7 and 8.   The loop shown was traced at a 60 Hz repetition rate.

Fig. 7 shows the behavior of the rf sideband effect as a function of this additional static field. In this data the static field was applied in a direction perpendicular to the rf field which was 8 Oe at 85 MHz for all cases except the top control spectrum. It is seen that the static field in fact has very little effect on the formation of rf sidebands. There is a slight decrease in sideband production for H_o = 1000 Oe, however there is actually a slight increase in the rf effect for H_o = 320 Oe, where as we have seen there are only a very few domains and domain walls in the sample.

THE MAGNETOSTRICTIVE MODEL

The experiments just discussed already rule out several possible mechanisms for the sideband effect. Ruled out for example is the entire class of models which envoke the motion of ferromagnetic domains walls such as the domain-wall passage model (11) proposed by Perlow.

One model for the magneto acoustic coupling which is consistent with all of the sideband properties discussed so far is the model of magnetostriction. Magnetostriction is the change in shape of a ferromagnetic sample in response to changes in the magnetization. The changes in shape arise because the principal term in the Hamiltonian for the exchange coupling between neighboring atomic magnetic moments has a dipole character,

$$f(r)(\cos^2\varphi - \tfrac{1}{3}),$$

depending on both the separation r and on the angle that the aligned moments make with respect to r (12). Thus exchange forces tend to distort the crystal lattice. The distortion of the lattice proceeds until an equilibrium is reached with the opposing elastic restoring forces of the strained lattice.

It is a characteristic of magnetostriction, because of the $\cos^2\varphi$ dependence of the exchange coupling, that large changes in lattice shape occur during rotations of the domain magnetization, but that movement of 180° domain walls has little magnetostrictive effect. On this basis one can see why the large $\vec{H}_o$ field perpendicular to the perturbing field

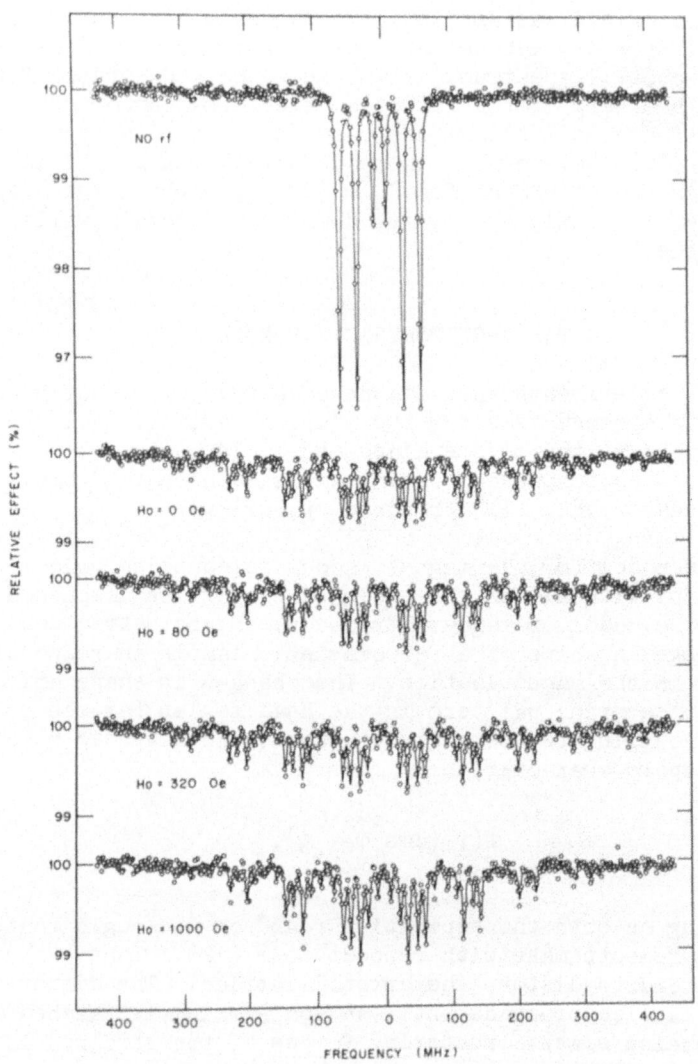

Fig. 7   The dependence of the rf sideband effect on additional applied field $\vec{H}_o$ oriented perpendicular to $\vec{H}_{rf}$.
The rf field for each of the lower four spectra had a peak
amplitude of 8 Oe at 85 MHz.

$\vec{H}_{rf}$ of Fig. 7 has only a small effect on sideband formation. When $\vec{H}_0$ and $\vec{H}_{rf}$ are perpendicular their vector sum is a time-varying rotating field oscillating about $\vec{H}_0$ in the plane of the sample foil. This large rotating applied field causes the sample magnetization to rotate even more easily than it would if $\vec{H}_0$ were zero, thus accounting for the slightly larger sideband effect observed with $\vec{H}_0$ fields of 80 and 320 Oe. When the $\vec{H}_0$ field is increased to 1000 Oe, however, the angle through which the resultant vector sum oscillates becomes somewhat smaller. This tends to slightly reduce the magnetostrictive distortion and therefore according to the model also the sideband effect.

If on the other hand $\vec{H}_0$ were to be aligned parallel to $\vec{H}_{rf}$, then the possibilities for rotation of the domain magnetization become substantially reduced as the sample approaches magnetic saturation along $\vec{H}_0$. Thus the magneto-striction model predicts that the sideband effect will decrease rapidly with increasing $\vec{H}_0$ provided $\vec{H}_0$ and $\vec{H}_{rf}$ are parallel. This in fact is what happens experimentally as shown in Fig. 8.

In view of this rather good qualitative agreement it is appropriate to ask whether the magnetostriction effect is large enough to account for the formation of rf sidebands. Several other authors (11,13) have considered this question and have decided that magnetostrictive effects are in fact too small. I will propose a possible way out of this diffi-culty in the discussion to follow.

We will make an order of magnitude calculation of the rf magnetostrictive effect assuming that the ferromagnetic material is isotropic and that the tabulated constants of magnetostriction are also appropriate at radio frequencies. Under these assumptions the fractional change in length due to magnetostriction can be written (12) as

$$\frac{\Delta \ell}{\ell} = \frac{3}{2} \Lambda \left(\frac{\vec{M}}{\vec{M}_s} - \frac{1}{3} \right)$$

where Λ is the magnetostrictive constant, M is the magnet-ization and M_s is the saturated magnetization for the sample.

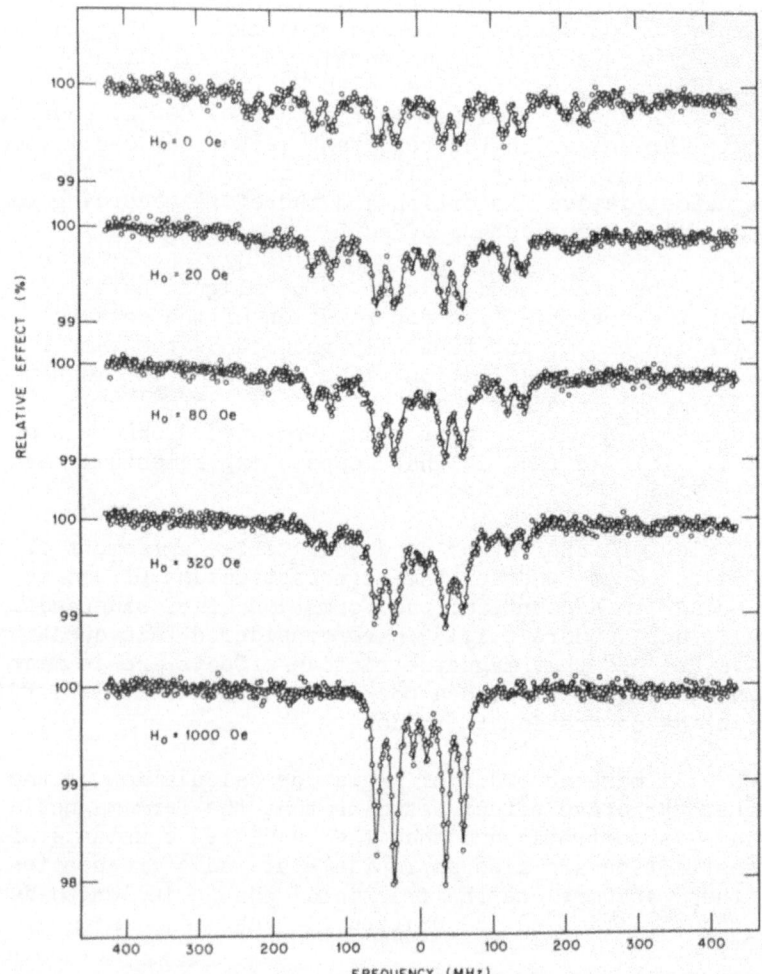

Fig. 8   The dependence of the rf sideband effect on an additional applied field $\vec{H}_o$ oriented parallel to $\vec{H}_{rf}$. The rf field for all spectra in the figure had a peak amplitude of 8 Oe at 88 MHz.

Now if a small rf field $\vec{H}_{rf}$ of frequency ω_m is applied to the sample it will induce a time dependence to the magnetization,

$$\vec{M}(t) = \vec{M}_{STATIC} + \vec{M}_{rf}e^{i\omega_m t} ,$$

In our magnetostrictive sample this time varying magnetization leads to a time dependent change in sample shape,

$$\Delta\ell(t) = \frac{3}{2}\ell \Lambda \left[\frac{2M_{STATIC}M_{rf}}{M_s^2} e^{i\omega_m t} + \frac{M_{rf}^2}{M_s^2} e^{i2\omega_m t} \right] , \qquad (5)$$

which is the expression for rf magnetostriction that we are seeking. The expression tells us that magnetostrictive oscillations will be generated at both the driving frequency ω_m and at $2\omega_m$, but it should be remembered that in general $M_{rf} \ll M_{STATIC}$ so that the harmonic term is usually negligable or very small. We may now relate the magnetostriction model to our earlier sideband formalism by noting that $\Delta\ell(t)$ corresponds approximately to the Doppler amplitude $x(t)$. Thus for this model the calculation of the modulation index of equation (4) becomes

$$m^2 = \frac{x_o^2}{\bar{\lambda}^2} \approx \frac{1}{\bar{\lambda}^2}\left(3\ell \Lambda \frac{M_{STATIC}M_{rf}}{M_s^2} \right)^2 \qquad (6)$$

Proceeding now to the actual numerical evaluation of equation (6) we note that

$$M_{rf} \leq \mu_{ROT}H_{rf}, \qquad (7)$$

where μ_{ROT} is the initial permeability of the sample due only to rotation (14) of the magnetization. Also for our purposes it is sufficiently accurate to replace M_{STATIC}/M_s by 1 since $M_{STATIC} \leq M_s$. For iron metal (14) $\mu_{ROT} \approx 29$ G/Oe, $M_s = 21.5$ kG, and $\Lambda = 20\times10^{-6}$. Using the experimental conditions obtaining for the sideband data of Fig. 2 gives $H_{rf} = 7.5$ Oe with a disk-shaped iron foil absorber 25 μm thick by 1.3 cm in diameter.

The remaining unspecified parameter is ℓ the sample
dimension before the rf is applied. To form Mössbauer sidebands the rf Doppler motion of the absorber nuclei must be
along the gamma ray axis - that is transverse to the plane
of the absorber foil. For this reason the first thought is
to assume that the proper choice for ℓ is one-half the foil
thickness. With this choice only magnetostrictive changes
in the foil thickness have importance, because presumably
it is only this motion that is along the gamma ray axis.
Substituting this 12.5 μm half-thickness together with the
other numerical values into equation (6), however, results
in a very small Doppler amplitude of 0.073 Å corresponding
to the modulation index $m^2 \leq 0.28$. A least-squares computer
fit to the sidebands in Fig. 2 using expression (4) gives
experimental modulation indices in the range between 0.5 and
18 (see Fig. 10). Thus our calculation of the magnetostriction mechanism produces a result which is at least an order
of magnitude too small. The discrepancy becomes even more
apparent for thinner foils. Substituting the half thickness
of an 8 μm iron foil into equation (6) yields a calculated
modulation index of 0.03, whereas the experimental index
needed to fit the data from these foils is again in the
order of 10.

Other authors (11,13) have obtained similar results
and for this reason have abandoned (11) the model.

It is possible, however, to get large indices from the
model by allowing ℓ to lie in the plane of the absorber foil,
since large changes in length are produced in that plane.
If the entire 1.3 cm diameter sample magnetostrictively re-
sponded with the proper phase to the rf, for example, the
outer edge of the foil would move 40 Å with respect to the
center producing enormous effective modulation indices, but
transverse to the gamma rays. The problem with the magnetostriction model thus is not that the Doppler motion is too
small but that the motion is not along the required axis.

I wish therefore to propose the following refinements
to the magnetostriction model.
1) The rf field produces magnetostrictive vibrations whose
amplitudes are initially in the plane of the absorber foil.
2) These rf vibrations propagate within the foil and
scatter off of grain boundaries, surfaces, and crystal dis-
locations with the result that some of the vibrational amplitudes come to have components along the gamma ray axis.

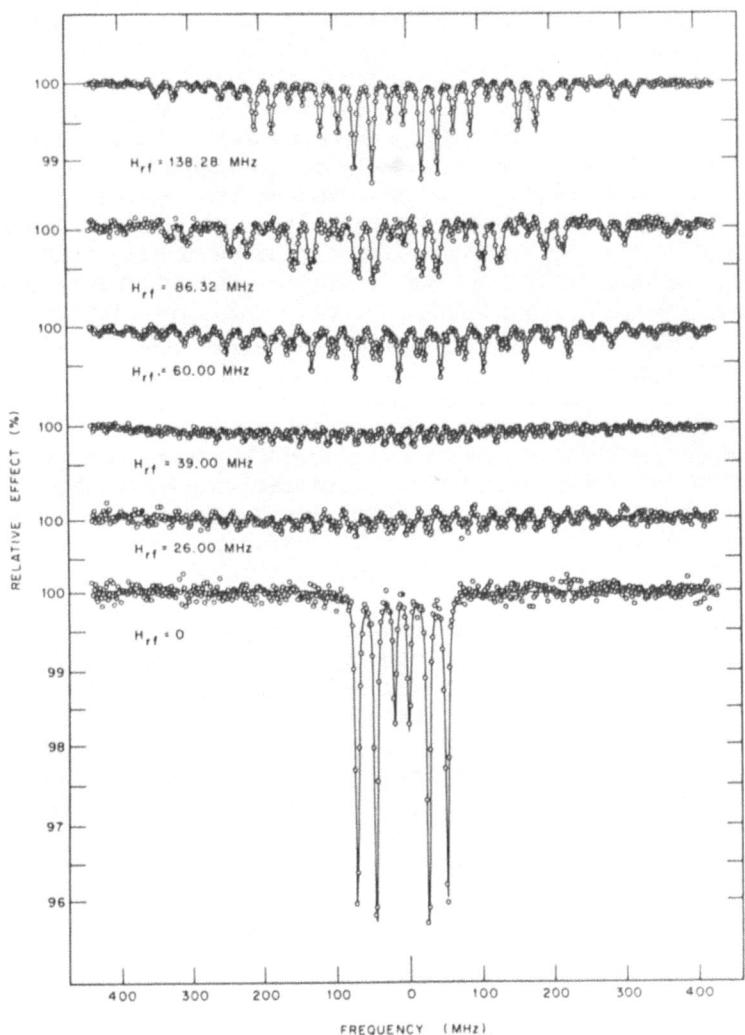

Fig. 9   The rf Mössbauer sideband effect as a function of
driving frequency. For each Mössbauer spectrum the rf
magnetic field was 7.5 Oe in peak amplitude directed in the
plane of the absorber. The absorber was a foil of 99.99%
purity natural iron 8 μm thick and 12.5 mm in diameter, which
had been previously annealed in dry hydrogen for several hours.

The detailed picture of sideband formation implied by this new model is clearly quite complex. As an example one would expect that the effectiveness of magnetostriction in producing rf vibrations within the foil would vary depending on the rf skin depth and on the possible existence of induced spin waves (15) which could effectively increase that depth. Also the distribution of the vibrational amplitudes of the magnetostrictive motion might be expected to depend on the communication of the relative acoustic phases between different parts of the sample area. This rf acoustic phase information should in turn depend on the velocity of rf sound in the material. Finally the proposed internal propagation and scattering of the acoustic vibrations should almost surely result in a complex redistribution of the vibrational amplitudes.

Even though the details are complex the new model does provide an explicit physical picture of the magneto-acoustic coupling. A picture which is consistent with all known experimental data, which is sufficiently detailed to inspire further study, and most importantly which produces acoustic vibrations of the proper amplitude to account for sidebands.

Before completing the numerical calculation with this new model, however, we need to examine more closely the experimental data showing the dependence of the sidebands on the driving frequency. Part of the data we want is in Fig. 2. There the peak amplitude of the rf field was held constant at 7.5 Oe and the frequency was varied by nearly an order of magnitude. Similar frequency data again using a 7.5 Oe driving field, but this time with an iron foil 8 μm thick are shown in Fig. 9.

The smooth curves through the data points in the spectra of both figures are least-squares computer fits with the relative sideband intensity coefficients as fitting parameters. The experimental values thus determined were then compared with computer generated tables of the sideband coefficient function given by expression (4). For each spectrum it was always possible to find a single modulation index $(x_0/\hbar)^2$ which when put into expression (4) would generate the experimental sideband intensity coefficients to within about 20%. In this way the sideband effect for each spectrum could be characterized by a single number, the modulation index.

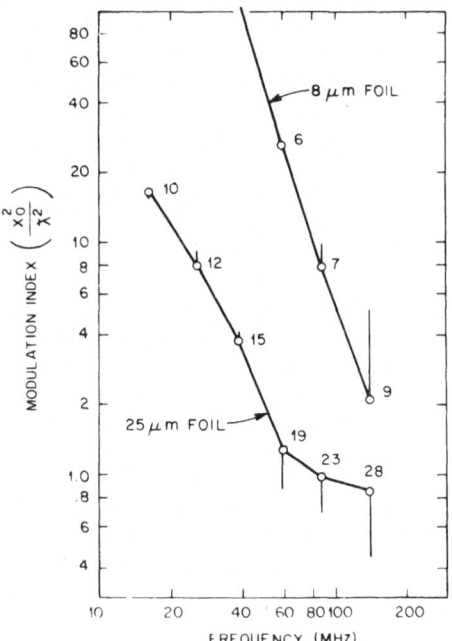

Fig. 10   Experimentally derived sideband modulation index parameters plotted as a function of driving frequency.   The modulation indices plotted here are derived from the data of Figs. 2 and 9 by the procedure discussed in the text.

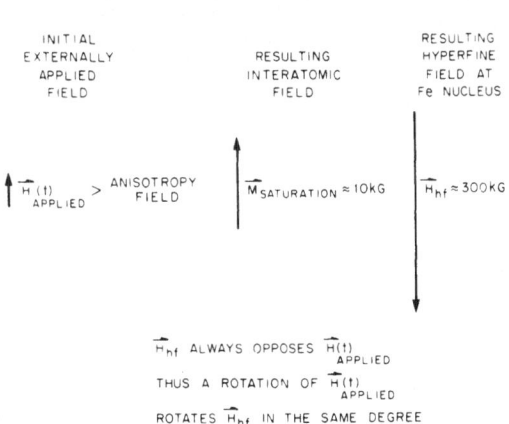

Fig. 11   The enhancement mechanism proposed to account for the rf collapse effect.

Fig. 10 is a plot of these experimentally derived
modulation indices obtained from the frequency data of
Figures 2 and 9. The circle is the modulation index that
gave the best fit to the zeroth order sideband coefficient,
and the vertical line extending away from it is the range
of indices required to overlap all of the other orders of
sideband coefficients. The length of this line then is
simply a measure of how well or poorly expression (4) actually does characterize the experimental sideband coefficients.

The first thing about this data that deserves comment
is the strong frequency dependence of the effect. Over the
rf range examined there does not appear to be any resonant
behavior, but the experimental modulation indices drop mono-
tonicly by several orders of magnitude as the driving
frequency is increased. Empirically the modulation indices
seem to vary as one over the driving frequency raised to the
3rd or 3.5 power.

The other striking characteristic of the data is the
apparent lack of importance of the rf skin depth for the
effect. The rf skin depth

$$d = \frac{1}{2\pi}\left(\frac{\rho}{\mu\nu}\right)^{\frac{1}{2}} \tag{8}$$

varies substantially over the range of frequency shown in
Fig. 10 so that the 25 μm iron foil which is 20 skin depths
thick at 16 MHz becomes 56 skin depths thick at 138 MHz.
For reference the calculated number of rf skin depths to
the center of the foil is printed near each modulation
index in Fig. 10. In looking at these numbers it is well
to remember that in a distance of only 2 skin depths the rf
field is reduced to 13% of its original value. One would
therefore expect the domain magnetization to rotate following the applied rf only in the extreme surface regions of
the foil. This would appear to require that all of the
significant magnetostrictive motion also occurs only in the
thin skin depth region near the surface.

But the Mössbauer technique is sensitive to all of the
Fe^{57} nuclei throughout the sample volume, not just those in
the surface regions where the rf magnetostrictive vibrations
presumably originate. Further, from the size of the rf
sideband effect observed it is clear that a substantial
fraction of all of these nuclei are vibrating at the applied

frequency. Thus the second proposal of our refined magnetostriction model is given experimental support; the propagation of the acoustic vibrations to the inner part of the
foil is clearly required if all of the significant magnetostriction occurs in the skin depth region, and yet the
Doppler vibrations are felt throughout the sample volume.

The interpretation of the rf skin depth is somewhat
complicated, however, by the possibility that spin waves
might be generated in our iron foil by the rf field (15).
Spin waves would also result in a kind of propagation from
the surface region to the sample interior. But the quantity carried by the spin waves is just the changing magnetization induced by the applied rf field. From this it
appears that the only contribution that spin waves can make
to sideband formation is the possibility that they would
effectively increase the rf penetration and thus the apparent rf skin depth of the sample.

We will conclude our discussion of the rf Mössbauer
sideband effect with some still tentative ideas on the role
of the acoustic propagation velocity in sideband formation.
Consider the area of an iron foil responding magnetostrictively to an rf field. Suppose the phase of the rf is such
that the foil is contracting in the direction along the
instantaneous $\vec{H}_{rf}$ vector. The contraction is the elastic
response of the iron crystallites to a change in the domain
magnetization, and thus proceeds at the velocity of sound in
the material. Because of this low velocity the contraction
will have hardly begun before the rf phase of the magnetization will have reversed, whereupon the elastic forces
will initiate an expansion. It would seem from this that
these elastic responses will be underlined coherent only over the small
region through which the acoustic wave propagates during an
rf field reversal.

Further, since our discussion of magnetostrictive vibration amplitudes implicitly assumed a coherent response,
it would seem that the proper value of ℓ that one should
use in equation (6) is just the radius of this virtual circle
of coherent acoustic amplitude. This radius will be on the
order of v/ν. For iron $v = 5 \times 10^5$ cm/sec which for
$\nu = 100$ MHz gives $\ell = 50$ μm. When this value of ℓ is substituted in equation (6) with the other numerical values as
before, we obtain a modulation index on the order of 4.5
which is in very good agreement with the experimental
values at 100 MHz in Fig. 10.

Furthermore, this coherence concept suggests the possible reason for the characteristic rolloff of the modulation index with increasing frequency. As we have seen the coherence length ℓ goes as $1/\nu$, so that from equation (6) $m^2 \propto 1/\nu$. But the number of atoms put into a given state of vibration by the magnetostriction is also proportional to ℓ and thus to $1/\nu$. Finally as we have said earlier unless spin wave phenomena nullify its effect, the rf skin depth should also be important in defining the number of atoms put into magnetostrictive vibration. But skin depth (eq. 8) goes as $1/\nu^{\frac{1}{2}}$, so that putting all of these factors together, we see the modulation index goes as

$$m^2 \propto \frac{1}{\nu}^{3.5} .$$

As was pointed out earlier this is in fact the approximate experimental dependence that we observed in Fig. 10.

THE rf COLLAPSE EFFECT

The other rf effect that I wish to discuss is the rf induced collapse of the hyperfine field (2). In searching for this effect I was able to use the same experimental equipment (Fig. 1) that is used in rf sideband experiments. The required rf fields are now however larger, and the absorber foils are more specialized.

The mechanism of this new effect is extreme ferromagnetic enhancement of the rf field. What this means is most easily explained with the help of Fig. 11. Reading Fig. 11 from left to right imagine for a moment that it is a static case and that we have applied to our ferromagnetic foil sample a field pointing up which is larger than the anistropy field. As a result of this applied field the ferromagnetic foil will become saturated or nearly so. If it is an iron foil it will have a saturation magnetization of 21.5 kG; for permalloy it will be of the order of 10 kG. The resulting hyperfine field induced at an iron nucleus in either of these materials will, because of the core polarization of the iron s electrons, point in the opposite or downwards direction with a strength typically of the order of 300 kG.

The result of all of this is that the 300 kG iron hyperfine field has become constrained to oppose the relatively small applied field. The applied field of only a few tens of oersteds is thus able to control the direction of the much larger hyperfine field. If the sense of the applied field is made to oscillate through the angle π in time, then the much larger hyperfine field at the iron nuclei will be forced by this enhancement mechanism to follow a similar directional oscillation.

The physics of this becomes interesting if the hyperfine field can be made to reverse its direction back and forth in a time comparable to the time for a nuclear Larmor precession. The nuclear Larmor frequency ν_L is of course just the usual precession frequency of the magnetic moment $\vec{\mu}$ of a nucleus about the normally static hyperfine field,

$$\nu_L = \frac{\vec{\mu} \cdot \vec{H}_{rf}}{Ih} .$$

But if the hyperfine field itself is rapidly changing direction at, for example, a rate several times the Larmor precession frequency, then the hyperfine field ceases to define a unique axis of magnetic energy quantization and the nuclear moment $\vec{\mu}$ will experience a zero field averaged over the relatively long nuclear lifetime of interest.

The experimental signature that this rf induced zero field condition has occurred in a Fe^{57} Mössbauer spectrum would be quite dramatic. The entire six-line absorption spectrum characteristic of iron in a magnetic environment should collapse together to become a single central line. This of course is the reason we have named the effect rf collapse.

Before proceeding further it is well to satisfy oneself that the internal magnetization actually will follow the reversal oscillations of the applied rf field at very high frequencies. The quantity of interest to us then is the switching time τ required in order to reverse the magnetic flux in our sample in response to a reversal of the driving field. The switching times of many magnetic materials have been measured (16,17,18) by workers interested in applications to digital computers. The times for reversal of the flux have been found to depend both on the material itself and on the peak amplitude of the driving field. Typical data (18)

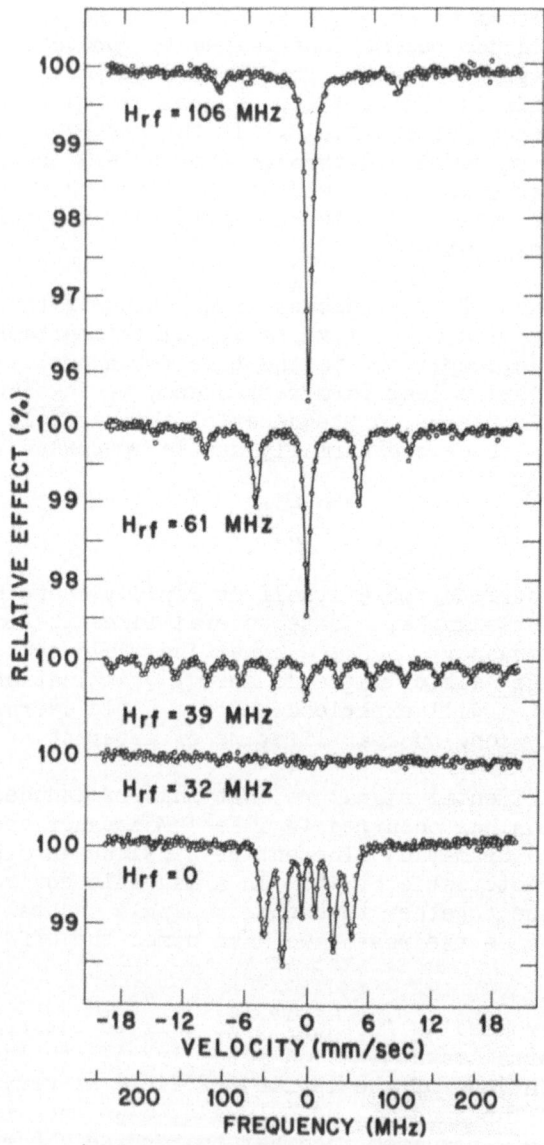

Fig. 12 The rf collapse effect as a function of driving
frequency. The absorber was a 6 μm thick foil of
58%Fe - 42%Ni permalloy annealed for 5 hours at 1035°C. An
rf field of 15 Oe peak amplitude was applied in the plane
of the absorber for each of the top four spectra.

for 80%Ni-20%Fe permalloy which has been studied most shows
that the reversal times decrease very sharply for increasing
driving field, with τ decreasing from 250nsec at 4 Oe to
27nsec at 6 Oe and finally dropping to less than 10nsec at
7.3 Oe. If one uses this data (18) to extrapolate to still
higher driving fields one obtains estimated reversal times
of the order of 1nsec for 10 Oe driving fields.

The conclusion therefore is that if we have the proper
magnetic sample and if we use driving fields $\vec{H}_{rf}$ in excess
of 10 Oe, then the internal magnetization of our sample will
indeed follow reversals of $\vec{H}_{rf}$ up to at least 100 MHz. Now
the Larmor precession frequency for an Fe^{57} nucleus in
permalloy at 400°K is about 21 MHz. This means if our rf
equipment is operating at 100 MHz that the entire hyperfine
field of ~270 kG will reverse itself nearly five times during
the time it takes the Fe^{57} nuclear moment to make one precession around the hyperfine field. As a result the time
average hyperfine field should be zero, and we should see
the entire six-line hyperfine spectrum of permalloy collapse
to a single central Mössbauer absorption line.

The experiment was done using a reasonably iron-rich
permalloy alloy, 58%Fe - 42%Ni, in the form of an absorber
foil 6 μm thick and 12.5 mm in diameter. The foil was annealed in dry flowing hydrogen gas at 1035°C for 5 hours and
then quickly quench-cooled at a rate of 600°C per minute.
Experimental results with this annealed permalloy absorber
foil are shown in Fig. 12.

The bottom spectrum in the figure is the control taken
with no rf field on the permalloy abosrber. We see the usual
six-line Mössbauer absorption pattern which is characteristic
of the Fe^{57} nucleus in a magnetic environment. Each of the
other four spectra in the figure were taken under the same
conditions except that for these spectra an rf field of
15 Oe peak amplitude was applied in the plane of the permalloy
foil.

In the top spectrum the reversal frequency of the 15 Oe
rf field was 106 MHz. This 106 MHz data shows the rf collapse
effect very clearly. The resolved six-line permalloy hyperfine pattern has disappeared and has been replaced by a
single dominant absorption line at zero velocity, just as
the rf collapse model predicts. Even the area of the

collapsed spectrum is consistent with the combined areas of
the six hyperfine lines in the bottom spectrum.

Besides the collapsed line at zero velocity, however,
there are in addition first order rf sidebands in this
spectrum displaced ±106 MHz from the central line. The
sidebands also show collapse single line structure. This
appearance of rf sidebands in our collapse spectrum is not
at all surprising. The 58%Fe - 42%Ni permalloy alloy that
we used is after all magnetostrictive so that the appearance
of sidebands in this experiment is perfectly consistent with
and in fact required by the modified magnetostriction model
for rf sidebands that we discussed earlier.

The simultaneous appearance of both the collapse and
sideband effects in the same experimental spectrum turns out
to be a useful check on the interpretation of the collapse.
This is because rf sidebands can be generated only in ferro-
magnetically ordered samples; (see Figs. 3 and 4 in the sideband section and the accompanying discussion). Thus the
collapsed sidebands at 106 MHz on either side of the central
line assure us that the permalloy foil was indeed ferro-
magnetically ordered, that is below the Curie temperature,
during the rf collapse experiment.

This points up a rather interesting physical situation
which occurs in our sample during rf collapse. The sideband
effect is after all an atomic effect in the sense that the
magnetostrictive mechanism operates on an atomic or even
many-atom microcrystalline scale. The collapse effect in
contrast requires only that the hyperfine field of each individual and isolated nucleus reverse direction rapidly with
respect to the Larmor precession time of that nucleus. The
existence of the sideband effect shows that on an atomic
scale the permalloy is magnetically ordered, that is that
there are multiatomic magnetostrictive interactions. The
existence of the collapse effect on the other hand shows
that the same sample at the same time on the nuclear scale
is magnetically disordered, having a zero effective field as
a result of the rapid rf reversals.

This disorder induced by the rf field on the nuclear
scale is in some ways similar to the usual lack of magnetic
order seen for a ferromagnet heated above its Curie temperature. All such samples at a temperature above T_c of course
also show no magnetic hyperfine field. But in this thermal

case it is believed that the thermal agitation of the lattice
produces random movements in the magnetization direction,
and that the core polarization interaction of the s electrons
transfers this random motion to the hyperfine field.

Thus from the point of view of an iron nucleus the rf
induced field reversals and the thermally induced random
field motion are similar in effect. In each case what is
believed to be a large hyperfine field is effectively time-
averaged to zero by directional fluctuations of the field.
The normal temperature induced fluctuation effect is random
in both time and spacial direction, whereas the rf induced
fluctuations are completely coherent in both space and time.
But such differences become all the more interesting when
one realizes that the analogies between rf collapse experiments and the corresponding effects associated with magnetic
order-disorder transitions are otherwise so close for a
Mössbauer nucleus.

In view of this analogy between rf induced and thermally induced collapse the dependence of the rf collapse on
driving frequency takes on special interest. How will the
collapsed single line which we observe at 106 MHz degrade
as the frequency is lowered?

Only a preliminary answer to this important question is
at hand; it is contained in the remaining spectra of Fig. 12.
As the driving frequency is lowered from 106 MHz to 61 and
then to 39 MHz one finds almost no change in the rf collapse
effect, but a large increase in the rf sideband effect.
This increase in the modulation index for sidebands as the
frequency is lowered is of course predicted by the magnetostriction model and is consistent with the behavior that we
have already seen in Fig. 10.

As the driving frequency is lowered still further to
32 MHz, however, a very dramatic change occurs in our spectra.
All effects seem to be washed out in the statistical noise.
We know that there is no discontinuous behavior in the rf
sideband effect over this frequency range, so apparently the
washing-out is due primarily to a change in the rf collapse
effect.

What we are seeing is presumably the degradation of the
collapse effect that we were looking for. One of course
expects on the basis of the collapse model that at some

point as the driving frequency approaches the nuclear Larmor
frequency, the induced reversals of the hyperfine field will
be too infrequent to completely cancel the hyperfine field.
This incomplete cancellation of the hyperfine field could
well result in a broadened partially collapsed pattern, and
this broadened pattern when replicated many times at the
32 MHz sideband spacing would produce the apparently washed-
out spectrum.

As pointed out above these are preliminary results.
It is intended to pursue this frequency dependence study in
considerably more detail. By studying the collapse effect
in the frequency regions where the averaging of the hyperfine field is incomplete, one should for the first time be
able to simulate thermal relaxation processes with a precisely controlled frequency of coherent field reversals.

As a further check of the rf collapse model and as a
way of exploring further, it was decided to study the collapse effect in the presence of an additional static magnetic
field $\vec{H}_O$ directed parallel to $\vec{H}_{rf}$, the 15 Oe rf field. The
motivation for these experiments was as follows. If the
amplitude of $\vec{H}_O$ is small compared to the amplitude of $\vec{H}_{rf}$,
then because the vector sum of these fields is again a field
oscillating through the angle π one would expect little change
in the enhancement mechanism and therefore little change in
the behavior of the collapse with respect to the $H_O=0$ condition. In contrast if H_O is larger than H_{rf} the superposi-
tion of the two fields will produce an oscillating applied
field, but one which is always pointing in the same direction
with only the magnitude changing. For such an applied field
both the magnetization and the hyperfine field should be
stationary to first order, so one would expect <u>no</u> collapse
effect to occur.

This prediction is beautifully confirmed in the experimental data of Figs. 13 and 14. These data were taken with
various H_O fields but always the same H_{rf} field which was
approximately 15 Oe in peak amplitude at 106 MHz. As shown
in Fig. 13 all of the data taken with $H_O \lesssim H_{rf}$ show the rf
collapse effect in varying degrees, but in agreement with
our model the $H_O=20$ Oe spectrum shows no rf collapse since
in this case the bias field is significantly larger than the
oscillating field. Fig. 14 is merely this $H_O=20$ Oe spectrum
replotted with an expanded vertical scale. With this
expanded scale it is easy to see in detail the entire

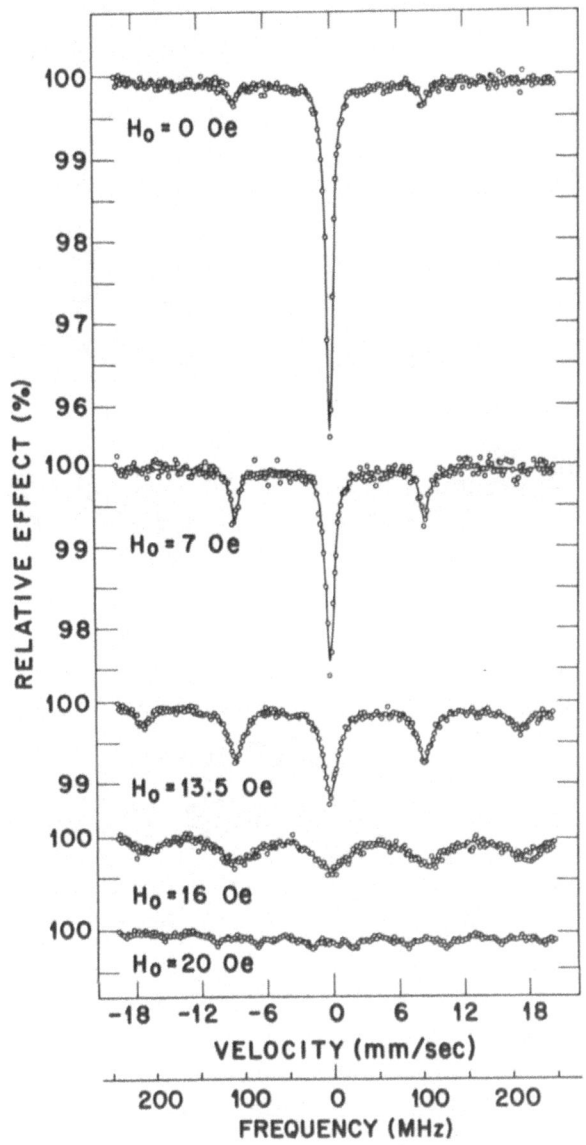

Fig. 13   The dependence of the rf collapse effect on an additional applied field $\vec{H}_0$ oriented parallel to $\vec{H}_{rf}$. The rf field for all of the spectra in the figure had a peak amplitude of 15 Oe at 106 MHz.

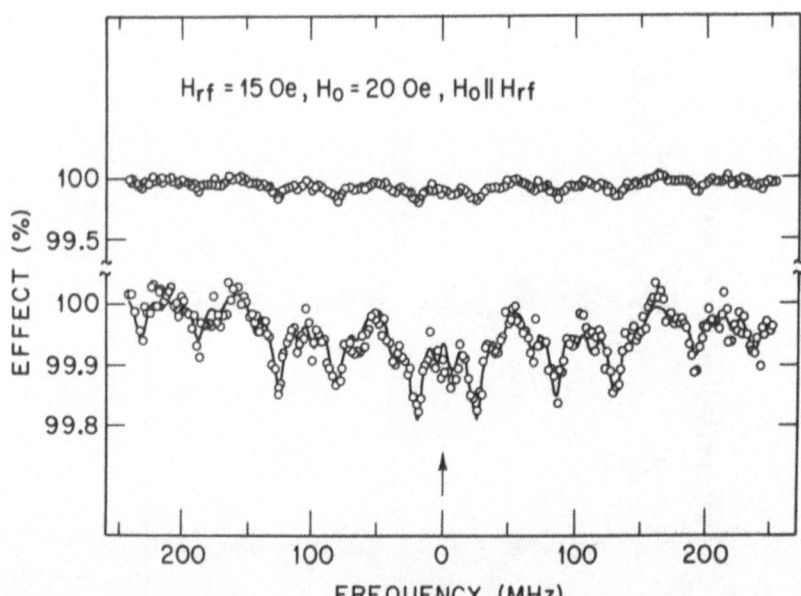

Fig. 14   The H_o = 20 Oe spectrum of Fig. 13 replotted on an
expanded vertical scale. The upper trace is the H_o = 20 Oe
spectrum as it appeared in Fig. 13. The lower trace is the
same spectrum replotted with the vertical scale expanded by
five times. The smooth curve which can be seen in the expanded plot is a least squares fit to the data assuming a
noncollapsed six-line spectrum accompanied by sidebands at
multiples of 106 MHz.

noncollapsed six-line hyperfine spectrum and its associated 106 MHz sideband companions.

In conclusion the rf collapse effect appears to be understandable in terms of the simple model that we have presented. The effect seems to hold out great promise for further research. A particularly promising avenue for further work appears to be the study of rf controlled hyperfine relaxation effects.

ACKNOWLEDGMENTS

The author would like to thank Drs. J. Dillon and G. Chin for helpful discussions, and Messrs. Charles Lichtenwalner, John Eimess and William Black for help of a technical nature.

REFERENCES

1. N. D. Heiman, Loren Pfeiffer and J. C. Walker, Phys. Rev. Lett. 21, 93 (1968).
2. Loren Pfeiffer, J. Appl. Phys. 42, 1725 (1971).
3. A. C. Gossard, A. M. Portis and W. J. Sandle, Phys. and Chem. Solids 17, (1961).
4. S. L. Ruby and D. I. Bolef, Phys. Rev. Lett. 5, 5 (1960).
5. T. E. Cranshaw and P. Reivari, Proc. Phys. Soc. (London) 90, 1059 (1967).
6. T. Mishovy and D. I. Bolef, in Mössbauer Effect Methodology, Vol. 4, pg. 13-35, I. J. Gruverman ed., Plenum Press, 1968.
7. Loren Pfeiffer, N. D. Heiman and J. C. Walker (to be published).
8. G. N. Watson, "A Treatise on the Theory of Bessel Functions", Cambridge Univ. Press 1966.
9. A. Abragam, "L'Effet Mössbauer", pg. 24, Gordon and Breach Publishers, New York, 1964.
10. M. Eibschütz, L. Pfeiffer and J. W. Nielson, J. Appl. Phys. 41 1276 (1970).
11. G. J. Perlow, Phys. Rev. 172, 319 (1968).
12. S. Chikazumi, Physics of Magnetism, Chapt. 8 (John Wiley and Sons, Inc., N.Y. 1969).
13. G. Asti, G. Albanese and C. Bucci, Phys. Rev. 184, 260 (1969).

14. S. Chikazumi, Physics of Magnetism, pg. 263 (John Wiley and Sons, Inc., N.Y. 1969).

15. The author is indebted to V. Jaccarino for pointing out the possible importance of induce spin waves in the sideband problem.

16. E. M. Gyorgy, J. Appl. Phys. 315, 110S (1960).

17. M. H. Kryder and F. B. Humphrey, J. Appl. Phys. 41, 1130 (1970).

18. F. B. Humphrey, J. Applied Phys. 35, 911 (1964).

THE DEBYE INTEGRALS, THE THERMAL SHIFT

AND THE MOSSBAUER FRACTION

Juergen Heberle

Department of Physics and Astronomy

State University of New York at Buffalo

Buffalo, New York 14214

If a Debye spectrum is assumed for the frequencies of lattice vibrations, the theoretical expressions for the Mössbauer fraction and for the thermal shift contain Debye integrals. It is shown how these integrals can be evaluated easily by means of infinite series. Only a few terms of these series provide adequate accuracy. A short numerical table is presented.

INTRODUCTION

In certain applications of the Debye model to problems of the solid state, one encounters integrals of the form

$$(1) \qquad D_p(z) = \int_0^z \frac{u^p \, du}{e^u - 1}$$

where $p = 1, 3$, or 4. We call such an integral a *Debye integral.*

Originally such an integral arose in the Debye theory of specific heat, according to which the internal energy of a crystal lattice containing N atoms is given by

$$(2) \qquad U = \frac{9}{8} Nk\theta + 9NkT \left(\frac{T}{\theta}\right)^3 D_3(\theta/T) \quad ,$$

where the first term represents the zero-point

299

energy.   It was found by Pound and Rebka [1], and
also by Josephson [2], that the *thermal shift* of
the Mössbauer resonance is related to the   average
lattice energy per atom.   Accordingly, we have for
the   thermal   shift   (expressed   as   a   Doppler
velocity)

(3) $S(T) = S(0) + \dfrac{9kT}{2mc}\left(\dfrac{T}{\theta}\right)^3 D_3(\theta/T)$.

For the *Mössbauer fraction*   (also   called   the
recoil-free fraction) the formula

(4) $f = \exp\left\{-\dfrac{6R}{k\theta}\left[\dfrac{1}{4} + \left(\dfrac{T}{\theta}\right)^2 D_1(\theta/T)\right]\right\}$

has been given by Mössbauer and Wiedemann [3].   A
detailed derivation of formula (4) has been worked
out   by   Kaufman   and   Lipkin [4].   The connection
between the Mössbauer fraction and the temperature
factor of Debye and Waller was also pointed out by
Tzara and Barlatoud [5].

Hardy, Parker   and   Walker   [6]   have   called
attention   to the need for a way to evaluate $D_1(z)$
on a digital computer by a method   that   does   not
entail   the slow process of numerical integration.
They have found a simple approximation

(5) $D_1(z) \approx 1.6449 \; [1 - \exp(-0.64486\, z)]$

and discuss its accuracy.   They have also given an
error function which improves the accuracy.

Contrary   to   widespread   belief,   these
integrals   can be evaluated without resorting to a
numerical-integration   method   such   as   Simpson's
rule.   It is our purpose to call attention to the
existence of two infinite series that   permit   the
evaluation of the Debye integrals for any value of
p   to   any   desired degree of accuracy.   It turns
out that only a few terms   of   these   series   will
yield   a   result whose accuracy exceeds the likely
needs of experimentalists.   The reader may wish to
consult the treatise of Knopp [7,8] or a   handbook
[9,10] for the pertinent mathematical background.

THE BERNOULLI NUMBERS

As we shall see below, for small values of z, the Debye integrals can be expressed in terms of the Bernoulli numbers B_n . These interesting numbers are often defined by means of the equation

$$(6) \qquad \frac{u}{e^u - 1} = \sum_{n=0}^{\infty} B_n \frac{u^n}{n!} .$$

We develop here those properties of the Bernoulli numbers that are relevant to our purpose.

It follows from (6) that

$$(7a) \qquad B_0 = 1 , \qquad B_1 = -1/2 ,$$

$$(7b) \qquad B_{2n+1} = 0 \qquad \text{for} \qquad n \geqslant 1.$$

Other Bernoulli numbers can be computed by successive applications of the formula [Knopp, Sec. 105]

$$(8) \qquad \sum_{r=0}^{n-1} \binom{n}{r} B_r = 0 .$$

We transform equation (8) into

$$(9) \qquad B_{2n} = \frac{1}{2} - \frac{1}{2n+1} \sum_{l=0}^{n-1} \binom{2n+1}{2l} B_{2l} .$$

From equation (9) one obtains

$$B_2 = 1/6 , \qquad B_4 = -1/30 \quad \text{and} \quad B_6 = 1/42.$$

Another useful form is

$$(10) \qquad B_{2n} = (-1)^{n-1} \frac{2(2n)!}{(2\pi)^{2n}} \left(1 + \frac{1}{2^{2n}} + \frac{1}{3^{2n}} + \cdots \right)$$

which brings out the important fact that the B_{2n} alternate in sign.

INFINITE SERIES

There are two series, one for low temperatures, the other for high values of T. First we treat the low-temperature expansion. We

begin by making use of the Riemann zeta function
[Ref. 9, p. 807]

$$\int_0^\infty \frac{u^p du}{e^u - 1} = p! \, \zeta(p+1) \, .$$

Then equation (1) becomes

$$D_p(z) = p! \, \zeta(p+1) - \int_z^\infty \frac{u^p du}{e^u - 1} \, .$$

We expand the denominator of the integrand

$$\int_z^\infty \frac{u^p e^{-u} du}{1 - e^{-u}} = \sum_{n=1}^\infty \int_z^\infty u^p e^{-nu} du \, .$$

It is readily confirmed by differentiation that
the indefinite integral yields

$$\int u^p e^{-nu} du = \frac{-p!}{n} e^{-nu} \sum_{\ell=0}^p \frac{u^{p-\ell}}{(p-\ell)! \, n^\ell} \, .$$

Now we put all the pieces together and obtain

$$D_p(z) = p! \left[\zeta(p+1) - z^p \sum_{n=1}^\infty \frac{x^n}{n} \sum_{\ell=0}^p \frac{1}{(p-\ell)!(nz)^\ell} \right]$$

where we have introduced the abbreviation

$$x = \exp(-z) \, .$$

The terms with $\ell = 0$ can be summed by means of

$$\log(1-x) = - \sum_{n=1}^\infty \frac{x^n}{n} \, .$$

The final form of this series is

$$(11) \quad D_p(z) = p! \, \zeta(p+1) + z^p \log(1-x) -$$

$$- p! \, z^p \sum_{n=1}^\infty \frac{x^n}{n} \sum_{\ell=1}^p \frac{1}{(p-\ell)!(nz)^\ell} \, .$$

In particular we have

$$(12) \quad D_1(z) = \frac{\pi^2}{6} + z \log(1-x) - \sum_{n=1}^\infty \frac{x^n}{n^2}$$

and

(13) $D_3(z) = \dfrac{\pi^4}{15} + z^3 \log(1-x) -$

$$- 3 \sum_{n=1}^{\infty} \frac{x^n}{n^4} (n^2 z^2 + 2nz + 2) \cdot$$

Although the above expansion is valid for all positive values of z, its convergence is rather slow when z is small, i.e. at high temperatures. We proceed to the case of small values of z and begin by using equation (6)

$$D_p(z) = \int_c^z \sum_{n=0}^{\infty} \frac{B_n}{n!} u^{n+p-1} \, du \cdot$$

This is readily integrated, and we obtain

(14) $$D_p(z) = z^p \sum_{n=0}^{\infty} \frac{B_n z^n}{n!\,(n+p)} \cdot$$

In the literature one meets also the Debye functions which are related to our integrals by

$$\Phi_p(z) = \frac{p}{z^p} D_p(z) \cdot$$

Then one has

$$\Phi_p(0) = 1, \qquad \Phi_p(\infty) = 0 .$$

The Debye functions have been tabulated for $p = 1,2,3$ and 4 [Ref. 9, p. 998].

E R R O R   E S T I M A T E S

Let us investigate the effect of terminating the infinite series in (11) at $n = m$. We write

$$S = S_m + R_m$$

where

$$S_m = \sum_{n=1}^{m} T_n$$

and

$$R_m = \sum_{n=m+1}^{\infty} T_n = \sum_{n=m}^{\infty} T_n - T_m$$

and

$$T_n = \frac{x^n}{n} \sum_{\ell=1}^{p} \frac{1}{(p-\ell)!\,(n\,z)^\ell}$$

$$= \frac{x^m}{m}\,\frac{m}{n}\,x^{n-m} \sum_{\ell=1}^{p} \frac{(m/n)^\ell}{(p-\ell)!\,(m\,z)^\ell} \quad.$$

We see that for $n > m$

$$T_n < \frac{x^m}{m}\,x^{n-m} \sum_{\ell=1}^{p} \frac{1}{(p-\ell)!\,(m\,z)^\ell} = T_m\,x^{n-m}.$$

It follows that

$$R_m + T_m < T_m \sum_{n=m}^{\infty} x^{n-m} = T_m \sum_{r=0}^{\infty} x^r = \frac{T_m}{1-x} \quad.$$

or that $\quad R_m < \dfrac{x}{1-x}\,T_m \quad.$

Thus we are able to conclude that

$$(15) \quad 0 < [\,p!\,\zeta(p+1) + z^p \log(1-x) - p!\,z^p S_m\,] -$$
$$- D_p(z) < p!\,z^p\,T_m\,\frac{x}{1-x} \quad.$$

The inequality (15) shows that the error incurred by terminating the series S is less than the quantity on the right-hand side of (15).

The next question is that of the convergence of the series in equation (14). When n is greater than unity, the terms in odd n vanish. The ratio of the 2(n+1)th term to the (2n)th term is obtained with the aid of (10) as

$$(-1) \left(\frac{z}{2\pi}\right)^2 \frac{1}{1 + \dfrac{2}{2n+p}} \frac{1 + \dfrac{1}{2^{2(n+1)}} + \cdots}{1 + \dfrac{1}{2^{2n}} + \cdots} \, .$$

From the behavior of this ratio, we conclude that the series (14) converges only for $|z| < 2\pi$. The error incurred by terminating the series will be less than the next term.

The actual errors are, of course, smaller than indicated by the above estimates. In fact, with just a few terms one can attain an accuracy better than most users are likely to require. Our computations have shown that the following formulas give results with an error less than 0.1 percent on the specified ranges:

$$D_1(z) = z \left(1 - \frac{z}{4} + \frac{z^2}{36} - \frac{z^4}{3600}\right) \qquad (0 \leq z \leq 2.2)$$

$$D_1(z) = \frac{\pi^2}{6} + z \log(1-x) - x - (x/2)^2 \qquad (1.7 \leq z < \infty)$$

$$D_3(z) = \frac{z^3}{3}\left(1 - \frac{3}{8}z + \frac{z^2}{20} - \frac{z^4}{1680} + \frac{z^6}{90720}\right) \qquad (0 \leq z \leq 2.5)$$

$$D_3(z) = \frac{\pi^4}{15} + z^3 \log(1-x) - \qquad (2.5 \leq z < \infty)$$

$$- 3 \sum_{n=1}^{2} \frac{x^4}{n^4}(n^2 z^2 + 2nz + 2)$$

NUMERICAL VALUES

An extensive table of $D_1(z)$ and of related quantities has been presented by Muir [11]. We have used our series to recompute some of Muir's values and have found no serious discrepancies. The largest disagreement (0.000 002) occurred at $z = 20$. We believe that the combination of computers and these series has rendered such tables somewhat obsolete. A table, however, is

still useful for checking a computer program; we
provide a short table of $D_1(z)$ and $D_3(z)$ for this
purpose.

z	$D_1(z)$	$D_3(z)$
0.1	0.097 527 8	0.000 321 0
0.5	0.440 963 6	0.034 373 5
1.0	0.777 504 6	0.224 805 2
2.0	1.213 894 6	1.176 342 6
3.0	1.441 305 7	2.552 218 5
4.0	1.552 592 1	3.877 054 2
5.0	1.604 381 0	4.899 892 2
6.0	1.627 562 8	5.585 855 4
7.0	1.637 635 9	6.003 169 0
8.0	1.641 914 4	6.239 623 8
9.0	1.643 699 9	6.366 573 9
10.0	1.644 434 7	6.431 921 9
20.0	1.644 934 0	6.493 920 2
∞	1.644 934 1	6.493 939 4

C O N C L U S I O N

The results expressed in equations (11), (13)
and   (14)   have already been stated, without
derivation, in a somewhat different but equivalent
form by C. Herring [12] and I. Stegun [13].

For very low temperatures, one can take

$$D_1(z) \approx \frac{\pi^2}{6}$$

which upon substitution in equation (4) leads to

$$f \approx \exp\left\{ -\frac{3R}{2k\theta} \left[1 + \frac{2}{3} \left(\frac{\pi T}{\theta}\right)^2 \right] \right\} ,$$

a result that was stated, without derivation,
already in 1959 by Pound and Rebka [14].

Another   quantity of interest in lattice
dynamics is the mean-square velocity of an atom.

At sufficiently high temperatures, this is given by the classical formula

$$\langle v^2 \rangle = 3kT/m \quad .$$

By use of equation (14), one can see how this quantity deviates from the classical value in the case of a Debye solid:

$$\langle v^2 \rangle = (3kT/m) \ [1 + \frac{1}{20}\left(\frac{\theta}{T}\right)^2 - \frac{1}{1680}\left(\frac{\theta}{T}\right)^4 + \cdots] \quad .$$

This formula was already obtained earlier in connection with an experimental study [15] of the thermal shift of the Fe-57 Mössbauer resonance in various host metals.

ACKNOWLEDGEMENTS

The computations were carried out on the CDC-6400 computer at the Computing Center of the State University of New York at Buffalo, which is partially supported by the N.S.F. Grant GP7318. The author is grateful to the Research Foundation of the State University of New York for the award of a Faculty Research Fellowship which made it possible to begin this work.

REFERENCES

1. R. V. Pound and G. A. Rebka, Jr.,
    Phys. Rev. Letters 4:274(1960).

2. B. D. Josephson,
    Phys. Rev. Letters 4:341(1960).

3. R. L. Mössbauer and W. H. Wiedemann,
    Z. Physik, 159: 33(1960).

4. B. Kaufman and H. J. Lipkin,
    Ann. Phys. (N.Y.) 18:294(1962).

5. C. Tzara and R. Barlatoud,
    Phys. Rev. Letters 4:405(1960).

6.   K. A. Hardy, F. T. Parker and J. C. Walker,
      Nucl. Instr. Meth. <u>86</u>:171(1970).

7.   K. Knopp, Theorie und Anwendung der
      unendlichen Reihen, 5th edition
      (Springer-Verlag, Heidelberg, 1964).
8.   K. Knopp, Theory and Application of Infinite
      Series, (Blackie and Son, Ltd., London, 1951).

9.   Handbook of Mathematical Functions, edited by
      M. Abramowitz and I.A. Stegun (Dover, New
      York, 1965).

10.   Fundamental Formulas of Physics, edited
      by D. H. Menzel, (Dover, New York, 1960).

11.   A. H. Muir, Jr., "Tables and Graphs for
      Computing Debye-Waller Factors in Mössbauer
      Effect Studies", Report AI-6699, Atomics
      International, Canoga Park,1962.

12.   C. Herring in vol. 2 of Ref. 10 on p. 604.
13.   I. Stegun in Ref. 9 on p. 998 .

14.   R. V. Pound and G. A. Rebka, Jr.,
      Phys. Rev. Letters <u>3</u>:439(1959).

15.   J. P. Schiffer, P. N. Parks and J. Heberle,
      Phys. Rev. <u>133</u>: A1553(1964).